Poster Program 2024

Held at the 2024 AIChE Spring Meeting and 20th Global Congress on Process Safety

New Orleans, Louisiana, USA
24-28 March 2024

ISBN: 979-8-3313-1718-8

Printed from e-media with permission by:

Curran Associates, Inc.
57 Morehouse Lane
Red Hook, NY 12571

Printed with permission by Curran Associates, Inc. (2025)

For permission requests, please contact AIChE
at the address below.

AIChE
120 Wall Street, FL 23
New York, NY 10005-4020

Phone: (800) 242-4363
Fax: (203) 775-5177

www.aiche.org

Additional copies of this publication are available from:

Curran Associates, Inc.
57 Morehouse Lane
Red Hook, NY 12571 USA
Phone: 845-758-0400
Fax: 845-758-2633
Email: curran@proceedings.com
Web: www.proceedings.com

Poster Program 2024

Held at the 2024 AIChE Spring Meeting and 20th Global Congress on Process Safety

New Orleans, Louisiana, USA
24-28 March 2024

AIChE
The Global Home of Chemical Engineers

TABLE OF CONTENTS

POSTER SESSION

POSTER SESSION II

ML/AI IN MMS 1

ADVANCES IN MATERIALS SIMULATIONS

ML/AI IN MMS 2

ADVANCES IN MMS

BIOMOLECULAR SIMULATIONS

CRYSTALLIZATION

CATALYSIS + REACTION ENGINEERING

Dissipative Particle Dynamics Characterization of ultra-low crosslinked (ULC) ionic microgel pastes

E. Daniel Cárdenas-Vásquez, L. Andrew Lyon

There has been an increasing interest during the last decade in the use of colloidal microgels for drug delivery, tissue engineering and other biomedical applications. However, experimental approaches to obtaining a deep understanding of the physical and chemical properties of these complex and dynamic materials at very small time and length scales such as at the atomic and molecular levels can be extremely challenging. Here, we present a novel computational approach to understanding the effect of dissolved ions on the properties of ultra-low crosslinked (ULC) pastes. Using a bottom-top approach - Dissipative Particle Dynamics (DPD) - we perform simulations to understand the nature of the charges, charge displacement and net charge effects in the microgel-microgel interactions within the ULC pastes. We show that all-atom Molecular Dynamics simulations for the more important functional groups coupled with coarse-grained PDP yield results that allow us to visualize the nature of the particle-particle interactions. We emphasize the necessity for energy-dependent and independent experiments required to carefully account for the charge displacement in the surrounding ionic cloud of the ULCs. Finally, we share future ideas that for a meso-scale understanding of ULC pastes.

Materials Informatics for Structure-Property Relationships (MISPR) for Liquid Solutions

Rasha Atwi, Matthew Bliss, Nav Nidhi Rajput

In this poster, I will present a general purpose, community supported, open-source computational framework coined MISPR (Materials Informatics for Structure–Property Relationships) for guiding and accelerating materials discovery, optimization, and deployment of complex multicomponent liquid solutions. MISPR can tackle the twin challenges of automating and parallelizing multi-scale molecular simulations and developing machine learning frameworks for advanced scientific applications in chemistry, materials science and engineering. It is high-throughput multi-scale computational infrastructure that seamlessly integrates classical molecular dynamics (MD) simulations with density functional theory (DFT).[1] By enabling high-performance data analytics and coupling between different methods and scales, MISPR addresses critical challenges arising from the needs of automated workflow management and data provenance recording. The major features of MISPR include automated DFT and MD simulations, error handling, derivation of molecular and ensemble properties, and creation of output databases that organize results from individual calculations to enable reproducibility and transparency. I will describe newly developed fully automated DFT and MD workflows implemented in MISPR to compute various electronic properties such as nuclear magnetic resonance chemical shift, binding energy, bond dissociation energy, and redox potential with support for multiple methods such as electron transfer and proton-coupled electron transfer reactions.[2] The infrastructure also enables the characterization of large-scale ensemble properties by providing MD workflows that calculate a wide range of structural and dynamical properties in liquid solutions and at solid-liquid interfaces. At the backend, the infrastructure interfaces with the Gaussian [3] software which enables electronic structure calculations of chemical systems, and LAMMPS [4] open-source code for MD simulations. LAMMPS workflows allow the execution of MD simulations in different ensembles and analysis of the dumped trajectories for various dynamical and structural properties using another open-source codebase developed by our lab called **MDPropTools.**[5] **MDPropTools** is a powerful standalone in-house suite of Python-based post-processing routines, to perform statistical analysis of thermodynamic, structural, and dynamical properties of MD simulations of liquids and solid-liquid interfaces.[5] Using MISPR and MDPropTools, we published the first publicly available database, **ComBat** ~2,000 computed QC and MD properties for reported lithium sulfur battery electrolytes composed of solvents spanning 16 different chemical classes.[6, 7] Another open source database, **MICRO** consists of ~3000 computationally predicted properties for application in carbon dioxide electroreduction. Integrated machine learning models allow exploration of a larger chemical and parameter space at a significantly higher speed to establish structure-property relationships in multicomponent solutions. MISPR employs the methodologies of materials informatics to facilitate understanding and prediction of phenomenological structure-property relationships, which are crucial to designing novel optimal materials for numerous scientific applications and engineering technologies.

Transferable Polarizable Model Development for Aqueous and Crystalline Alkali-Halide Salts

Philippe B. Baron, Athanassios Panagiotopoulos

Electrolyte solutions are vital to the development of novel and groundbreaking technologies such as new batteries and carbon sequestration methods. Accurate theoretical models that can predict the collective properties of electrolyte solutions (solubilities, activity coefficients, transport coefficients, etc.) are highly desirable as they can guide the rational design of new technologies that rely on these substances. Due to the complexity of aqueous electrolytes, the most successful approach to this modeling challenge is through computer simulation methods such as molecular dynamics. Unfortunately, this approach is complicated by the tradeoff between accuracy/transferability and computational efficiency in the description of interatomic energies. We argue that effectively describing the aqueous and crystalline chemical potential across a wide range of temperatures and pressures is particularly important, as these values determine solubility and solution nonideality.[1,2] Also, it has been shown that computer simulation models with poor solubility predictions, also provide poor predictions of solution structure.[2] Efficient classical force fields have struggled to describe these properties, and newly emerging Neural Network potentials are very promising, but do not yet provide quantitative agreement with experiment for these collective properties.[1,3,4] Recently the development of more advanced polarizable molecular Force Fields (FFs) has shown promise in describing electrolyte chemical potentials at a relatively low computational cost.[5] This work is focused on achieving temperature transferability and greater accuracy in chemical potential predictions from atomistic simulations of simple electrolyte solutions (alkali halide salts dissolved in water). This is done by conducting a systematic optimization of state-of-art, drude-oscillator-based Force Fields directly to aqueous chemical potential data at different temperatures. The targeting of the temperature-dependence of the chemical potential is emphasized not only for greater experimental applicability of these models, but also due to its connection to partial molar enthalpies as well as other properties.[5] The calculation of electrolyte chemical potentials and solubilities from molecular simulations is very challenging and thus this work centers around the development of an efficient workflow to refine classical, polarizable models to better reproduce free energy data. In order to increase efficiency, we simulate these force fields in LAMMPS using an extended Lagrangian framework and develop refined free energy calculation approaches. Additionally, a Bayesian loss function is employed that can be optimized locally using configurational reweighting, greatly reducing the number of samples of the objective function needed.[6] We hope that this approach will produce better, more transferable alkali-halide force fields, as well as establish an efficient procedure for addressing such complex force field optimization scenarios.

Novel tools for studying crystallization pathways of shaped particles into complex crystals

Domagoj Fijan, Brandon Butler, Philipp W. A. Schoenhoefer, Thomas Waltmann, Charlotte Zhao, Maria Ward Rashidi, Joshua Anderson, Sharon Glotzer

Nucleation and crystallization processes are critical in diverse materials, such as colloidal, soft, molecular, metallic, and ionic crystals. While there's considerable understanding of nucleation in simple crystals, research on complex crystals' nucleation is limited due to the lack of suitable computational tools. We address three main areas for improvement: an order parameter that describes local structures with continuous point group symmetry, a new neighbor list for shaped particles, and a tool for efficiently detecting and recording nucleation events.

Complex crystals exhibit multiple local environments, and to truly understand the nucleation process on a lowest level would require us to somehow track the progression of symmetry elements for each of these environments. However, traditional order parameters such as Steinhardt's (SOP) and its variants merge multiple symmetries into a single value, limiting their effectiveness in detailed analysis. Our novel Point Group Order Parameter (PGOP) overcomes this limitation by providing a per-particle order parameter based on symmetry operations. PGOP compares a particle's bond order diagram (BOD) with a perfect, symmetrized BOD using the Pearson correlation coefficient. This quantifies how well the local environment aligns with specific symmetry operations, allowing distinct differentiation of local environments in complex crystals.

The accuracy of PGOP depends on identifying neighboring particles. Traditional methods like Voronoi tessellation and Solid Angle Nearest Neighbors (SANN) often fail in systems with shaped particles. To resolve this, we introduce Shadow Projection Overlap of Obstacles for Neighbor Exclusion (SPOONE). SPOONE identifies neighbors by casting rays and determining intersections (blocking), providing more accurate neighbor lists for shaped particles, and enhancing PGOP's effectiveness. SPOONE can be utilized as a traditional or weighted neighbor list by considering the amount of blockage (shadow). This makes it very suitable in use for order parameter computation.

To facilitate the automatic detection of nucleation events and manage the data-intensive nature of these simulations, we developed *dupin*, a Python tool for event detection in molecular simulation trajectories. *dupin* promises to automate detection of events offering new potential applications of machine learning in the field. *dupin* identifies changes using order parameters and employs change point detection methods to record events with high temporal resolution. Since *dupin* can work on-the-fly during the simulation, it can also serve to minimize data storage and improves simulation efficiency by selectively recording only significant events.

To showcase the utility of our newly developed tools, we present results from comparisons in traditional situations suitable for each tool. We contrast PGOP with SOP in both simple and complex crystals. SPOONE's effectiveness is demonstrated in systems of elongated shaped particles, with performance comparably effective for nearly spherical particles. Additionally, we explore the impact of different neighbor lists on PGOP and SOP by comparing the results using SPOONE versus Voronoi methods. Additionally, we highlight *dupin*'s ability to efficiently identify various events in soft matter simulations, including intricate nucleation processes and polymer collapses, for comprehensive post-simulation analysis. These tools collectively promise to advance the study of nucleation and crystallization, potentially opening new research opportunities in large-scale crystallization projects and enabling application of emerging...

Molecular Simulation of Supercapacitors

Xiaobo Lin, Shern R. Tee, Debra Debra J. Searle, Peter Cummings

Molecular dynamics (MD) simulations using the constant potential method (CPM) have been proven to be an essential technique for studying the charge storage and charging dynamics of *homogeneous* electrodes. We will provide an overview of recent work [1] on CPM MD *operando* simulations for charging of nanoporous carbide-derived carbon supercapacitors with four distinct electrolytes, including ionic liquid (IL), mixed IL-solvent, and solvent-in-salt electrolytes. However, the CPM cannot capture the distinct atomic nature of *heteroatomic* electrodes, such as MXenes, a class of two-dimensional inorganic compounds that consist of atomically thin layers of transition metal carbides. To overcome this limitation, we developed the heteroatomic constant potential method (HCPM) for MXene supercapacitors. The proposed HCPM model takes into account the electronegativities of different atoms in heteroatomic electrodes. A brief overview of results obtained with Li-TFSI/AN (lithium bis(trifluoromethane sulfonyl)imide / acetonitrile)-based solvent-in-salt electrolytes will be provided [2].

References

Lin, X., Tee, S. R., Searles, D. J. & Cummings, P. T., 'Molecular insights on optimizing nanoporous carbon-based supercapacitors with various electrolytes,' *Electrochim. Acta* **474,** 143500 (2024). DOI: 10.1016/j.electacta.2023.143500

Lin, X., Tee, S. R., Kent, P. R. C., Searles, D. J. & Cummings, P. T., "Development of heteroatomic constant potential method with application to MXene-based supercapacitors," J. Chem. Theory Comp., submitted (2023).

molGrouper: Enabling Efficient, Exhaustive Exploration of Chemical Space Using Group-Based Approaches

Kieran Nehil-Puleo, Nicholas Craven, Peter Cummings, Clare McCabe

The exploration of chemical space stands as a crucial endeavor across various scientific domains, particularly in the realms of material and drug discovery, where the determination of novel molecules holds paramount importance. However, prevailing methods, especially those reliant on deep learning-based generative models, encounter significant limitations stemming from their inefficient representation of molecules and stochastic nature. These generative methods typically rely on an atomistic graph or sequence-based representations of molecules. Atomistic representations often result in a non-physical and inefficient sampling of the expansive landscape of potential molecular structures. In the field of organic chemistry, the advantageous representation of molecules as combinations of functional groups is well established. In response to the challenges facing previous generation methods, and with the known utility of functional groups, we have devised an open-source software, molGrouper, that uses pre-specified chemical groups as fundamental building blocks for molecular structures. The fundamental data structure of our software is the GroupGraph which represents functional groups in a molecule as nodes, with bonds between groups represented as edges. The key functionalities of molGrouper are the exhaustive generation of a complete chemical space, the translation of members of the chemical space to molecular simulation engines, and the conversion of molecular structures into data-efficient machine learnable inputs. molGrouper addresses the aforementioned needs of the expanding cheminformatic industry and provides a general and transferable method to represent the growing space of molecules.

Ammonia's behavior within MFI zeolite nanosheets

Rammanish Singh, Nathan Wang, Henry Wolters, J. Ilja Siepmann, Daniela Kohen

The success of catalytic and separation processes within micro-porous and nano-porous materials depend on the appropriate behavior of the molecules involved and the materials themselves. Over the years, significant improvements in practical applications have resulted from a fundamental understanding of these materials.

MFI zeolite nanosheets appear to be promising candidates for ammonia purification, yet little is known about its chemical behavior. We have used first-principle molecular dynamics techniques to simulate ammonia's behavior within MFI zeolites with explicit surface silanols. Our work explores H-bonding and proton transfer within this system and demonstrates interesting reactivity.

Transport of Dilute Suspensions in Porous Media with Heterogeneous Particle-Media Attractions

Wilfred Kwabena Darko, Deepak Mangal, Jacinta C. Conrad, Jeremy C. Palmer

Porous media, which find widespread use in practical applications, naturally exhibit variations in their composition and surface charge. These variations create diverse physicochemical interactions between the material and the particles moving through it. In our study, we employed Stokesian dynamics simulations to investigate the impact of these heterogeneous interactions on the diffusion and dispersion of particles in ordered arrays of nanoposts. Our findings reveal that, under quiescent conditions, particle transport involves two mechanisms: diffusion through open spaces and intermittent hopping between attractive sites on different nanoposts. As the degree of heterogeneity of attractions increases, the latter mechanism becomes more dominant. This leads to greater complexity in particle trajectories, deviations from the typical Gaussian behavior in particle displacement patterns, and a reduction in long-time particle diffusion. Similarly, under flow conditions characterized by low Péclet numbers (Pe), heightened attraction heterogeneity causes temporary particle clustering near the nanoposts, resulting in a wider particle distribution and enhanced dispersion in the direction of flow. However, at high Pe values where advection predominates, the longitudinal dispersion coefficient remains unaffected by attraction heterogeneity and follows the well-established Taylor-Aris dispersion pattern. Our research sheds light on the ways in which heterogeneous physicochemical interactions can impact the transport of particles in complex three-dimensional porous materials.

Computational Insights into Copper-Catalyzed Biological Methane Mitigation

Joshua Peterson

Methane is a potent greenhouse gas (GHG) that is blanketing and heating our planet at alarming rates, accounting for nearly 30% of total global warming. Atmospheric methane's global warming potential is ~86 times that of CO_2 on a 20-year time scale, necessitating effective mitigation strategies. Nature's first responders to the GHG warming issue are methanotrophic bacteria, equipped with copper-dependent particulate methane monooxygenase (pMMO), an enzyme that selectively oxidizes methane to methanol under ambient conditions. Activating the inert C-H bonds in methane both selectively and at room temperature is a challenge that has not yet been achieved in catalyst design.

Our goal is to understand what limits the efficiency of methane oxidation in these microbes to guide the design of optimal methane capture systems. Coupling computational methods with experimental studies of metalloenzymatic processes and recent cryoEM data, we are probing the structure-activity properties of pMMO. Employing quantum mechanical/molecular mechanical (QM/MM) we are characterizing the oxygen activation, methane binding, and the oxidation mechanisms. Using QM methods to benchmark the copper interactions will be a central focus. Ultimately, we aim to help identify the site of activation, the role of the surrounding protein environment, and the properties that stabilize transition state intermediates.

Modeling, Screening, and Technoeconomic Evaluation of Nanoporous Materials for Boil-Off Gas Treatment during Intercontinental Transportation of Liquefied Natural Gas

Sunghyun Yoon, Seongbin Ga, Jinwoo Park, Haneul Mun, Inkyu Lee, Yongchul Chung

Intercontinental transportation of liquefied natural gas (LNG) relies on the energy-intensive re-liquefaction process to minimize the boil-off gas (BOG) losses during the trip. Previous research efforts have focused on improving and optimization of the existing process designs to treat the BOGs. In this work, we developed an innovative process based on nanoporous materials called high-pressure and low-temperature (HPLT) adsorption process for the BOG treatment. We perform high-throughput, multi-scale modeling campaign to discover high-performing nanoporous materials for the BOG treatment. Our analyses show the developed HPLT adsorption process is economical than the current state-of-the-art processes for 8-day and 13-day trips with annualized saving of \$1.5 – 2.0 million per ship.

Investigating the Electrode/Electrolyte Interface of Sodium-Glyme Battery Electrolytes Via Constant Potential Method

Dodangodage Ishara Senadheera, Orlando Carrillo-Bohorquez, Ernest Nachaki, Ryan Jorn, Daniel Kuroda, Revati Kumar

Rechargeable batteries based on sodium salts and glyme solvents hold promise as large-scale energy storage solutions, yet further improvements are necessary for their commercial viability. Enhancing the electrochemical performance of sodium ion batteries requires more than just understanding the chemistry of electrolytes and electrode materials; it demands a comprehensive grasp of the underlying mechanisms governing the electrode-electrolyte interface. Particularly, the decomposition reactions at these interfaces give rise to surface films. Previous research has shown that the solvation structure of cations in the electrolyte profoundly impacts the formation and properties of these surface films.

Conducting molecular dynamics simulations on an atomistic model of the interface, with electrolyte atoms situated alongside polarizable electrodes, will provide a realistic representation of the interfacial phenomena in a battery. Constant Potential Method (CPM) is such method that allow the polarization of the electrode by the electrolyte.

In this study, a molecular dynamics investigation, validated experimentally, explores a 1 M NaTFSI salt in glymes of varying lengths placed between two graphite electrodes under constant bias potential. The focus lies on delineating the solvation environment surrounding sodium ions at the electrode-electrolyte interface concerning glyme chain length and applied potential. Results reveal distinct interface features for the diglyme/TFSI system compared to triglyme/TFSI and tetraglyme/TFSI systems. Computational predictions align well with experimentally measured capacitance, and the dominant solvation structures elucidate the system's electrochemical stability, consistent with cyclic voltammetry characterization.

Self-Assembly of Non-Spherical Colloids Using Coarse-Grained Simulations and Machine Learning Approaches

Bahadir Rusen Argun, Antonia Statt

Modern synthesis methods have enabled the fabrication of colloids with various different shapes. In addition, plastic waste degrades to nano- and microplastics particles in the environment, also leading to irregular shaped colloidal particles. These particles, polluting our habitats have unknown implications for our health and habitats. Therefore, understanding self-assembly of these non-spherical particles crucial not only for many applications, like bottom-up fabrication of large ordered systems, but also for understanding the fate of microplastic particles in the aquatic environment.
To simulate the non-spherical colloids, we utilized the commonly used composite bead approach to represent the different shapes. We find that this approach is not only expensive as the complexity of the colloidal shape increases, due to inter-body distance calculations, but also cumbersome as it is not always obvious how the composite beads should interact with each other. Here, we will highlight the use of data-driven methods to accelerate our simulations. We trained neural networks to directly predict energy, forces, and torques between rigid non-spherical particles based on their configuration, bypassing the need for inter-bead distance calculations. We show that MD simulations performed with neural net predicted forces and torques can accurately reproduce the structure and kinetics of the traditional simulations with explicit distance calculations. This approach can yield significant computational speedup and can be applied to irregular shapes with any pair interaction, given sufficient training data.

Computer-Verified BET Analysis Using the Lean Theorem Prover

John Velkey, Colin Jones, Tomáš Skřivan, Tyler R. Josephson

Chemical engineering frequently requires calculations for solving problems. Because of the wide range of computational tools and methods available for performing calculations of this nature, inconsistencies between methodologies are common. For example, when adsorption datasets were sent to sixty-one different labs, there was wide variation in the corresponding BET surface area reported, despite the fact that determination of BET surface area is a well-known calculation [1]. Consider the utility then, of programs that perform calculations using functions that have provable mathematics, that are "bug free" as a result of their mathematical verification. To illustrate this concept, we use Lean 4 to write a computer-checked derivation of BET theory, and then link this derivation to functions that regress the BET parameters for a data set.

Lean 4 is a novel, multi-purpose programming language that also functions as an interactive theorem prover. We first use Lean as a theorem prover, and formally verify the derivation of the BET adsorption theory. This proof is first presented in [2], originally implemented in an earlier version of the Lean interactive theorem prover. Proofs in Lean 4 are not identical to executable computing programs, which can be separately defined as functions. Not all computing functions can have proofs written on them, if they are strictly typed to function with floating-point numbers. This is due to the disjoint relationship between Real numbers and floats. Reals can be the subject of proofs, but because they potentially have infinite decimal digits, they cannot be computed or displayed by computer systems. Instead, scientific computing uses floats to perform actual calculations, but proofs about floats are not the same as proofs about reals; even a + b - b = a is not true for all floats. To make verified scientific computing possible in Lean, we introduce a method using polymorphic functions to connect proofs about real-valued functions to computations using the same functions with float-valued inputs. We use the developed code to analyze datasets from [1] performing BET regression, and compare our formally-verified results to those previously reported.

[1] Osterrieth, Johannes W. M., et al. "How Reproducible Are Surface Areas Calculated from the BET Equation?" Advanced Materials, vol. 34, no. 27, 23 May 2022, p. 2201502, https://doi.org/10.1002/adma.202201502.

[2] Bobbin, Maxwell P, et al. "Formalizing Chemical Physics Using the Lean Theorem Prover." Digital Discovery, **3**, 16 Nov 2024, p 264-280, https://doi.org/10.1039/D3DD00077J.

Active Learning of Alchemical Adsorption Simulations; Towards a Universal Adsorption Model.

Etinosa Osaro, Jair Fajardo-Rojas, Gregory Cooper, Diego A. Gómez-Gualdrón, Yamil Colón

Adsorption is a fundamental process studied in materials science and engineering because it plays a critical role in various applications, including gas storage and separation. Understanding and predicting gas adsorption within porous materials demands comprehensive computational simulations that are often resource intensive, limiting the identification of promising materials. Active learning (AL) methods offer an effective strategy to reduce the computational burden by selectively acquiring critical data for model training. Metal-organic frameworks (MOFs) exhibit immense potential across various adsorption applications due to their porous structure and their modular nature, leading to diverse pore sizes and chemistry that serve as an ideal platform to develop adsorption models. Here, we demonstrate the efficacy of AL in predicting gas adsorption within MOFs using "alchemical" molecules and their interactions as surrogates for real molecules. We first applied AL separately to each MOF, reducing the training dataset size by 57.5% while retaining predictive accuracy. Subsequently, we amalgamated the refined datasets across 1800 MOFs to train a multilayer perceptron (MLP) model, successfully predicting adsorption of real molecules. Furthermore, by integrating MOF features into the AL framework using principal component analysis (PCA), we navigated MOF space effectively, achieving high predictive accuracy with only a subset of MOFs. Our results highlight AL's efficiency in reducing dataset size, enhancing model performance, and offering insights into adsorption phenomenon in large datasets of MOFs. This study underscores AL's crucial role in advancing computational material science and developing more accurate and less data intensive models for gas adsorption in porous materials.

Zeonet: 3D Convolutional Neural Networks for Predicting Long-Chain Hydrocarbon Adsorption in Nanoporous Zeolites

Yachan Liu, Gustavo Perez, Zezhou Cheng, Aaron Sun, Ping Yang, Sam Hoover, Wei Fan, Subhransu Maji, Peng Bai

Zeolites, nanoporous materials widely utilized in the chemical industry, hold considerable promise for applications in chemical separations and catalysis. This study introduces ZeoNet, a representation learning framework employing 3D convolutional neural networks (ConvNets) and a volumetric representation based on distance grids, to predict Henry's constants for the adsorption of long-chain hydrocarbon molecules in more than 330000 all-silica zeolites. The investigation systematically explores and optimizes various factors impacting model performance, including ConvNet architecture, linear dimension, and spatial resolution of volumetric grids, batch size, optimizer, and learning rate. The top-performing ZeoNet achieved an impressive correlation coefficient ($r^2 = 0.977$) and a mean-squared error (MSE = 3.8) in lnkH, corresponding to a minimal error of 9.3 kJ/mol in adsorption free energy. Furthermore, feature attribution analysis and visualization of trained models revealed that ConvNets predominantly base their predictions on accessible pore volume rather than the region occupied by framework atoms. We compared ZeoNet with other common representations, including 2D multi-view distance images, 3D multi-channel grids, point clouds based on atom coordinates and on surface sampling, and graph networks. We found that 3D volumetric grids vastly outperformed the other representations, and that 3D ResNet18 with single-channel distance grids achieve the best balance between computational efficiency and accuracy for all-silica zeolites. Finally, we explore transfer learning to predict the adsorption of five other hydrocarbons in a data-constrained setting. This study offers insights into the optimal utilization of ConvNets for representing extended materials systems and provides a suite of pretrained models applicable to scenarios with limited training data and multiple targets.

Exploring Chain Length and OH-Spacing Effects of Choline-Chloride Based Alkanediol Deep Eutectic Solvents Via Classical Molecular Dynamics.

Daniel Kim, Yong Zhang, Edward Maginn

Deep eutectic solvents have become increasingly popular due to their environmentally friendly nature and high tunability. In this work, the extensively studied DES ethaline (a 1:2 mixture of choline chloride with ethylene glycol) is compared to other alkanediol-based DESs with ChCl as the hydrogen bond acceptor molecule for a wide range of compositions. Classical molecular dynamics simulations have been validated against experimental data and used to investigate the local liquid structure and dynamics that result from the hydrogen bond network of the solvent. It has been found that vicinal diol hydroxyl (OH) groups cause stronger intermolecular interactions than terminal OH groups. The viscosity of the solvent has also been found to correspond to the strength of the key interactions, or hydrogen bonds, of the system with the most significant being the Cl-OH and OH-OH self-interactions. Both alkyl chain length and OH spacing are important factors in DES design.

From Molecular Fragments to Novel High-Performing Fluids: The Discovery of Green Refrigerants.

Barnabas Agbodekhe, Edward Maginn

The Kigali Agreement of 2016 requires that environmentally harmful hydrofluorocarbons (HFCs), which are predominantly used as refrigerants but cause global warming, be phased out. The phase-out of current refrigerant compounds requires the design of green alternatives. Previous works in green refrigerant discovery have relied mainly on database screening in which it was reported that options for pure-component green refrigerants were very limited.

This work presents a novel and rigorous approach to the discovery of new materials applied to refrigerant discovery. We developed a computational molecule generation tool written in Python called FineSMILES. FineSMILES collects packets of molecular fragments made up of pre-selected elements based on an understanding of the problem and literature. These packets of molecular fragments are screened using structural and chemical constraints to identify packets that could be assembled into chemically feasible molecules. FineSMILES assembles these packets into complete molecules with outputs as SMILES strings. Application of the FineSMILES code generated hundreds of thousands of molecules, of which more than fifty percent are new molecules that are not present in the PubChem database of compounds used in previous refrigerant discovery projects. A key implication is that a new world of chemical compounds not previously explored for use as green refrigerants has been discovered. We then applied a simple and novel approach that integrates group contribution models and Sigma profiles with Gaussian process regression models for accurate and reliable property prediction. The predicted properties were then used to screen the refrigerant molecules for technical, environmental, and safety performance. Based on a combined consideration of predicted thermophysical, environmental, and safety properties, tens of molecules were identified as high-potential green refrigerants, with several of them not previously reported in the open literature. Future work includes further in-depth screening of the identified molecules using molecular simulations and quantum mechanical calculations.

Predicting Molecular Scale Dynamics and Kinetics Occurring during Photovoltaic Encapsulant Degradation

Mark Wilson, Hannah Dedmon, Jennifer Braid, Michael E. Chandross, Jessica Kustas

A predictive materials-based understanding of aging degradation in polymeric encapsulants is critical for achieving a 50-year lifespan for solar photovoltaic modules. Novel encapsulant materials are emerging, with the market share heading towards increasingly cost-effective choices. In most cases, a molecular scale understanding of the chemical degradation of these new materials is lacking, where many factors including chemical kinetics and diffusion of reaction products can simultaneously drive the aging process. A more complete understanding of the heterogeneous degradation of encapsulants requires knowledge of the migration of species, their concentration dependence, and their combined effects on degradation rates. In this newly funded DuraMAT project (FY24-FY25), we aim to use a combined experimental aging and computational approach to identify these degradation species and quantify the factors that initiate and potentially drive degradation.

Here, we establish two key modeling approaches used in the computational thrust of the project, demonstrated on the current industry-standard material choice, polyethylene vinyl acetate (EVA) copolymer. We use molecular modeling to assess material-specific properties. Steered molecular dynamics along with reactive force fields are being employed to investigate energetic barriers for chemical degradation kinetics. Additionally, the diffusional properties of degradation species are assessed from all-atom simulations with varied chemical compositions of EVA. Results from these two simulation approaches can be both chemically specific input parameters for higher length scale models and indicators of expected experimental behavior for a range of encapsulant chemical compositions. Our broader vision in this effort is to demonstrate the materials-based approach necessary to establish a predictive reaction-diffusion framework for novel materials.

Sandia National Laboratories is a multimission laboratory managed and operated by National Technology & Engineering Solutions of Sandia, LLC, a wholly owned subsidiary of Honeywell International Inc., for the U.S. Department of Energy's National Nuclear Security Administration under contract DE-NA0003525.

Large Property Models: A New Generative Paradigm

Brett Savoie

Generative models for the inverse design of molecules with particular properties have been heavily hyped but have yet to demonstrate significant gains over machine learning augmented expert intuition. A major challenge of such models is their extremely limited accuracy in predicting properties in the data scarce regime. For example, activity data for a drug target or stability data for a material may only number in the tens to hundreds of samples, which is insufficient to learn an accurate property to structure probabilistic mapping. We've hypothesized that the property to structure mapping becomes unique when a sufficient number of properties are supplied to the models during training. This hypothesis has several important corollaries if true. It would imply that data scarce properties can be completely determined by a set of more accessible molecular properties. It would also imply that a generative model trained on multiple properties would exhibit an accuracy phase transition after achieving a sufficient size—a process analogous to what has been observed in the context of large language models. To this end, we've built the first transformers capable of directly imagining molecules given a property description, which we dub "large property models". A key ingredient is supplementing these models during training with relatively basic but abundant chemical property data. Colloquially, we describe this approach as teaching the models basic chemistry before tasking them with predicting species with highly specific properties. The large property model paradigm will be covered in the poster submitted for discussion along with the architectures and case studies that we've looked at thus far.

Advanced Featurization and Characterization of MOF Pores for Adsorption

Arun Gopalan, Kaihang Shi, Lev Sarkisov

State-of-the-art approaches to computational discovery of MOFs can be thought of in terms of three key components. 1) Representation: Atoms and bonds, textural properties etc. 2) Prediction: Molecular simulations, Machine Learning (ML) etc. 3) Structure sampler/generator: Databases like CoRE-MOFs, ToBaCCo etc. or, Genetic Algorithms, Bayesian Optimization etc. Representation is key to all the downstream steps. Unlike chemical properties, adsorption correlates better with the characteristics of the 3D confining environment, which is very difficult to represent efficiently. Hence, we present a novel concept of 'pore graphs', a compact yet comprehensive way to retain any cavity information while adaptable to various applications including materials informatics and downstream Machine Learning (ML).

The void space in a material is converted into a network of interconnected pockets using custom algorithms overlayed atop 3D Watershed image segmentation. The connectivity of two adjacent pockets depends on the pore constriction (window) between them and Periodic Boundary Conditions (PBC). Conceptually, pore graphs therefore decouple the local and global influences, while taking advantage of the inherent pore symmetries present in crystalline materials.

Breaking down the void space into constituent pockets allows us to get the most out of existing pore descriptors. To demonstrate, we'll showcase three key applications in this work,
- Detect pore windows (internal and periodic to the unit cell)
- Detect the types of pockets using ML clustering (of different in size, shape or energy)
- Matching pockets from a database of materials by energy, to fit small, rigid molecules closely.

A comprehensive open-source Python package 'Mofography' will be released on GitHub pre-packaged with tools for pore graphs construction, the above applications, visualization, and high-throughput screening.

A Multi-Scale Graph Network for Synthetic Peptide Property Prediction

Daniel Garzon, Camille Bilodeau

While synthetic peptides can solve problems ranging from treating antibiotic-resistant diseases to catalyzing new reactions, resource and time costs make experimentally exploring their massive design space intractable. Deep learning models are well-suited for exploring the vast synthetic peptide design space because they can predict molecular properties in a rapid and parallelizable fashion. We present a new, multiscale graph neural network architecture designed to leverage the natural modularity of peptides. Our methodology is advantageous over previous approaches in that it is capable of 1) learning structural relationships between amino acids to facilitate the incorporation of non-natural amino acids, and 2) capturing nonlinear peptide features including staples, cycles, and stereoisomers. We demonstrate our model's performance for two tasks: first, chromatographic retention time (for which large datasets are available), and second, anti-bacterial activity (for which only small datasets are available). This work enables deep learning for the prediction of synthetic peptides and will serve as a foundation for deep learning-guided discovery of synthetic peptides.

Advancing Coarse-Graining Methods Using Statistical Mechanical Principles and Machine Learning

Alexander J. Pak

Multiscale modeling and simulations can provide microscopic insights into emergent dynamical processes that span (i) multiple spatial and temporal scales difficult to study through experiments alone and (ii) many disciplines, including chemistry, biology, physics, and materials science. Bottom-up coarse-graining is one such technique where information from high-resolution (e.g., atomistic) models is used to derive low-resolution (e.g., coarse-grained) models following statistical mechanical principles. Over the past decade, many algorithms to systematically determine coarse-grained mappings and effective interactions have been proposed. However, we posit that these techniques have yet to realize their full potential due to limitations imposed by traditional pseudo-particle representations described by pairwise, nonbonded interactions. With the emergence of high-throughput data generation techniques and machine learning algorithms, new opportunities to improve the fidelity of coarse-grained models should be explored. In this poster, we describe recent efforts by our group to push the boundaries of bottom-up coarse-graining by incorporating strategies from the data science communities. We will present progress along two tracks, where we incorporate many-body physics to go beyond traditional force-field functional forms and particle-based representations. We demonstrate our methodologies through several examples of small proteins and discuss anticipated improvements for applications to more complex biomolecular and soft matter systems.

Learning to Trust Machine-Learning-Accelerated Simulations for Material and Chemical Predictions

Rebecca Lindsey, Yanjun Lyu, Benjamin Laubach

Elucidating atomistically-resolved insights into chemical and material evolution in complex and inherently multiscaled systems is a historically challenging problem, particularly in cases where characteristic spatiotemporal scales preclude use of highly predictive first-principles methods. However, the advent of machine-learned interatomic models (ML-IAM) has transformed the simulation landscape, enabling "quantum accurate" predictions on previously inaccessible scales. There are a multitude of ML-IAM and training strategies currently available to choose from and a manifold of success stories surrounding their application, but grand challenges exist surrounding training set generation, reproducibility, uncertainty quantification, interpretability, and the overall computational expense associated with model development. We will discuss recent developments to ChIMES, a unique, physics-informed machine-learned interatomic model and semi-autonomous fitting framework aimed at overcoming these challenges, including recent efforts to leverage the underlying model form for data analysis and uncertainty quantification.

Molecular Simulation Study of Gas Adsorption and Transport in Room Temperature Ionic Liquid Membranes Supported By Carbon Nanotubes

Gaurab Sarkar, Andrew J. Schultz, Jeffrey Errington

Development of robust strategies to capture carbon dioxide from flue gas emissions in power plants represents a key step in addressing factors that contribute to climate change. Membranes featuring room temperature ionic liquids (RTILs) have been identified as potentially effective media for capturing carbon dioxide. The very low vapor pressures and tunable nature of RTILs make them attractive media for membranes. The Yu group at the University at Buffalo has recently pursued an approach in which the RTIL is supported by a mesh of single-walled carbon nanotubes (SWCNT). The approach results in a membrane with nano-sized features. The mesh structure potentially provides enhanced mechanical, transport, and selectivity characteristics. In this work, we use molecular simulation tools to study gas adsorption and transport properties within mesh-based RTIL structures, and compare the performance of these systems to bulk RTIL systems. We first discuss generation of a molecular model that aims to capture the essential features of the mesh structure. We next describe the Monte Carlo simulation protocol developed to study gas adsorption. Specifically, we employ a strategy in which the temperature, pressure, numbers of molecules of RTIL and SWCNT, and the activities of relevant gases are fixed. The approach enables us to capture the adsorption characteristics of various model flue gases over a range of pressures and compositions. Finally, we present results for the adsorption and diffusion characteristics of select systems over a broad range of conditions. These characteristics are then linked to the manner in which the RTIL and gases organize within the nano-confined space.

Capturing Reactions of Polymer Pyrolysis Using Physical Experiments and Reactive Molecular Dynamics

Adam Dumas, Tim Mallo, Phil Westmoreland

Physical and simulation experiments are being used together to develop molecularly accurate models of polymer pyrolysis and its kinetics. Understanding the kinetics and mechanisms of polymer pyrolysis has far-reaching applications from plastic recycling to solid rocket fuels. The needed understanding must come from a combination of measurements and predictive molecular theory and modeling. Our previous development of Reactive Molecular Dynamics through the RxnMD code and RMDff force field has demonstrated this approach's qualitative and quantitative value [1]. Our present work extends and deepens that knowledge using LAMMPS/ReaxFF [2].

Our physical tools include thermogravimetric analysis / differential scanning calorimetry (TGA/DSC), flash pyrolysis, and gas chromatography (GC) and two-dimensional GC with time-of-flight mass spectrometry (GCxGC/ToFMS), which give overall kinetics of mass loss, heats of melting and reaction, and compositions of gas and condensable products. This combination of rate data and products is crucial to proposing pathways and establishing mechanisms. However, except by DSC, there is no detection of reactions that do not produce volatile products, such as scission of polymer chains into nonvolatile products.

Reactive Molecular Dynamics simulates such reactions, capturing both the bond breaking and the physical strains that lead to bond scission. Our previous work [1] has demonstrated the ability of Reactive Molecular Dynamics (through RMDff/RxnMD) to predict bond-breaking activation energies for high-density polyethylene (HDPE). The simulations showed that increasing chain length caused a decrease in activation energies, likely due to potential energy in the form of strain from angular and torsional modes. This interesting conclusion cannot be captured by our current experimental techniques. RMDff also showed the ability to predict products successfully from HDPE decomposition.

Using LAMMPS/ReaxFF, we are extending this approach to polymers of different monomer complexity, including polyoxymethylene (POM) and poly (methyl methacrylate) (PMMA). We will present predicted bond-breaking activation energies and products, to be used in conjunction with our experimental data to develop detailed mechanisms for thermal pyrolysis of POM and PMMA.

References:

[1] K. D. Smith, M. Bruns, S. I. Stoliarov, M. R. Nyden, O. A. Ezekoye, and P. R. Westmoreland, "Assessing the effect of molecular weight on the kinetics of backbone scission reactions in polyethylene using Reactive Molecular Dynamics," *Polymer*, vol. 52, no. 14, pp. 3104–3111, Jun. 2011, doi: 10.1016/j.polymer.2011.04.035.

[2] Parallel reactive molecular dynamics: Numerical methods and algorithmic techniques, H. M. Aktulga, J. C. Fogarty, S. A. Pandit, A. Y. Grama, *Parallel Computing*, vol. 38, pp. 245-259, 2012, doi:10.1016/j.parco.2011.08.005

Development of Chimes Machine-Learned Interatomic Potentials for All Silica MFI Zeolites

Vallabh Vasudevan, Sayed Ahmad Almohri, Anakin Bahm, Rebecca Lindsey

Ammonia ranks as one of the most produced in the global chemical manufacturing, primarily used as a precursor in synthesizing fertilizers to satisfy the growing global food supply demands Currently, this demand is being met through the Haber-Bosch (HB) process, which is among the most energy intensive in the chemical industry. These energy demands can be largely attributed to the process used to separate the ammonia from the gaseous reactive mixture, i.e., which involves cooling the gas to 240 K to liquefy the ammonia. Current efforts to develop a more energy efficient industrial scale ammonia production process focus on developing improved catalysts and separation membranes to drive the reaction equilibrium. Nanoporous zeolite membranes hold great promise for application in this space. They have high thermal and chemical stabilities and show good selectivity and permeability for the separation of various mixtures, [1–4]. Spatial variation in adsorption, diffusion pathways and pore channel length control zeolite membrane performance, all of which are affected by the atomic arrangement of the zeolite framework. [5] An extensive understanding of the molecular mechanisms affecting adsorption, diffusion and selectivity and dynamic behavior of the framework are hard to approach using current experimental techniques. [5–7] Here, we will present a new machine-learned interatomic potential for bulk MFI zeolite based on the based on the ChIMES physics-informed machine-learned interatomic model and development framework. [8–14] We will highlight a new model training strategy that accounts for structural fluctuations by including strained and compressed structures. We will show that bulk structural properties predicted with our model such as Radial pair distribution functions and vibrational power spectra compare well with similar properties derived from DFT simulations. Our model can be used to simulate the zeolite pore channels and compare the predicted structural configurations with observed experimental structures derived from imaging techniques such as ptychography to better inform and predict atomic level framework dynamics and performance characteristics. [5] We will discuss possible strategies to extend our MFI zeolite model to other all-silica zeolite structures. Ultimately, our newly developed ChIMES machine-learned IAP can be used to provide useful insight into the role of zeolite as a separation media for HB process and to study the underlying interactions in zeolite-mediated catalytic processes.

References

[1] Hisao Inami et al. Membranes 11.5 (2021)

[2] Miguel Palomino et al. Langmuir 26.3 (2010)

[3] Jin Shang et al. Chem. Commun. 50 (35 2014)

[4] Jin Shang et al. J. Am. Chem. Soc. 134.46 (2012)

[5] Hui Zhang et al. Science 380.6645 (2023), pp. 633–638.

[6] Zhuoya Dong et al. J. Am. Chem. Soc. 145.12 (2023).

[7] Haozhi Sha et al. Science Advances 9.11 (2023)

[8] Rebecca K. Lindsey et al. J. Chem. Theory Comput. 13.12 (Dec. 2017).

[9] Rebecca K. Lindsey et al. The Journal of Chemical Physics 153.13 (2020)

[10] Rebecca K. Lindsey et al. J. Chem. Theory Comput. 15.1 (2019)

[12] ChIMES Calculator. URL: https://github.com/rk-lindsey/chimes_calculator.

[13] ChIMES Generator. URL: https://github.com/rk-lindsey/chimes_lsq.

[14] ChIMES Active Learning Driver. URL: https://github.com/rk-lindsey/al_driver.

Examining Structure and Dynamics in Polymers of Intrinsic Microporosity (PIMs) from Molecular Dynamics Simulations

Diana Aponte Claudio, Janani Sampath

Gas separation membranes play a crucial role in various industrial applications, offering notable advantages such as energy efficiency, cost-effectiveness, and versatility. Polymers of Intrinsic Microporosity (PIMs) have garnered significant attention due to their unique microporous structures and exceptional permeability-selectivity characteristics. The distribution of microporous voids in the polymer is known as free-volume element distribution, which significantly influences the passage of penetrants through the membrane. However, the performance of PIM-based membranes is often compromised by aging and plasticization phenomena, leading to reduced separation efficiency and overall effectiveness. Both aging and plasticization are strongly correlated with chain and backbone flexibility. Segmental dynamics exert a significant influence on these membrane relaxation phenomena. These dynamics occur at short lengths and time scales, presenting challenges for characterization using experimental techniques. Here, we employ molecular dynamics (MD) simulations to elucidate the origin and impact of local dynamics in PIM-based gas separation membranes. We aim to unravel the underlying mechanisms that compromise membrane performance by examining chain mobility and intermolecular interactions. Our approach begins with parameterizing PIMs systems using Ambertools, ensuring an accurate representation of their molecular structure. Then, we utilize the simulated polymerization algorithm, Polymatic, to generate and validate membrane structures, ensuring reproducibility and agreement with experimental properties. We explore various PIMs with different functionalities, including PIM-1, PIM-Py, and PIM-COOH, all of which are commercially available. By studying membranes with different functional groups, we seek to understand how these variations influence molecular structure, chain packing, free volume structure, and penetrant transport. Chain dynamics will be analyzed using the bond autocorrelation function, while free volume is computed using the void analysis kit VACUUMMS. The insights gained from our study will deepen our understanding of aging and plasticization phenomena in PIM-based membranes, shedding light on the roles of chain mobility and intermolecular interactions. Ultimately, this knowledge could inform the design of more robust and durable membrane materials, paving the way for developing the next generation of high-performance gas separation membranes.

Quantum Informed Machine Learning Potentials for Modeling Framework Flexibility in Nboffive-1-Ni Metal-Organic Framework

Xijun Wang, Hilal Daglar, Randall Snurr

Metal-organic frameworks (MOFs), with their distinctive porous structures and tunable chemical properties, have shown immense promise in the adsorption and separation of gases. Currently, the accurate simulation of their adsorptive properties remains a formidable challenge, especially for systems where the molecules fit very tightly in the pores. Traditional simulation methods often approximate the frameworks as rigid and do not account for the significant framework flexibility seen in materials such as NbOFFIVE-1-Ni.[1] This simplification may lead to significant deviations between simulated and experimental results, underscoring the critical need to incorporate framework flexibility into simulations. Ab initio molecular dynamics (AIMD) simulations offer the desired accuracy in modeling this flexibility but are limited by their extensive computational demands, rendering them impractical for prolonged simulations. Conversely, classical force field-based simulations offer computational feasibility but lack the necessary accuracy. To break this accuracy-efficiency trade-off, we have developed machine-learning force fields (ML FFs) trained on energies and forces from density functional theory to enhance MD simulations, accurately predicting the flexibility of the NbOFFIVE-1-Ni framework in the presence of water over nanosecond timescales. This innovative approach enables first-principles accuracy at the lower cost of classical force field-based methods. By combining MD simulations with the ML FF and force field-based grand canonical Monte Carlo (GCMC) methods, we can also predict water adsorption isotherms taking framework flexibility into account. These findings open up new options for MOF design and application in gas adsorption and separation.

References

[1] S. Kancharlapalli, R. Q. Snurr, "High-Throughput Screening of the CoRE-MOF-2019 Database for CO_2 Capture from Wet Flue Gas: A Multi-Scale Modeling Strategy", ACS Appl. Mater. Interfaces., **2023**, 15, 28084-28092.

An Data- and Information-Driven Approach for Computational Soft Materials Design

Ashley Guo

Controlling the emergence of order out of disorder is an overarching goal in the engineering of biological and soft materials. Conventional computational approaches have relied strongly on multi-scale molecular simulations aided by enhanced sampling techniques, including the recent proliferation of machine learning based methods. An exciting direction in the computational design of soft materials incorporates ideas from information theory into these approaches, drawing upon the connection between thermodynamic entropy and information entropy. This poster will highlight our recently-established research group's current interests at the intersection of molecular simulation, machine learning, and information theory, including new directions in developing polypeptide-based materials for contaminant capture in wastewater, as well as characterization of polymer phase separation using data compression algorithms.

Design of Peptides to Inhibit Multiple *Clostridium Difficile* toxin B Subtypes in the Presence of UDP-Glucose and Mn^{2+}

Corey Febo, Carol Hall

Clostridium difficile (C. diff) infections cause diarrhea and severe colitis. *C. diff* kills 30,000 out of the 500,000 people it affects every year in the United States. Current antibiotic treatments are not always effective due to antibiotic resistance. Short peptides are being considered as potential drug candidates to prevent and treat *C. diff* infection by inhibiting the biocatalytic activity of the two toxins, TcdA and TcdB, released by the bacterium. These toxins transfer glucose from UDP-Glucose to Rho/Rac family proteins, leading to the death of colonic epithelial cells. In our group's previous efforts, a solid-phase peptide screening library and the *Pep*tide *B*inding *D*esign (PepBD) algorithm were used to design peptides to prevent the binding of UDP-Glucose to TcdB. Interestingly, one of the most effective peptides identified, SB5, was found experimentally to bind to a different site than the one where UDP-Glucose binds. Instead, this peptide bound to the site on TcdB where the Rho/Rac protein binds, thus preventing cell death by preventing the Rho/Rac family proteins from being glycosylated. We aim to determine how SB5 binds to the TcdB1 and TcdB3 subtypes when UDP-Glucose and Mn^{2+} are already bound since UDP-Glucose and Mn^{2+} are both present on TcdB in cellular environments. This may help us to design peptides with a lower binding free energy to multiple TcdB subtypes than the SB5 peptide. We have applied a computational approach that combines explicit-solvent atomistic molecular dynamics simulations and the PepBD algorithm to design new and more-effective peptides for combating *C. diff* infections.

Playing Potato Head with Accelerated Sampling Methods for Fun and Profit: Alchemical Metadynamics and Replica Exchange of Expanded Ensembles

Wei-Tse Hsu, Anika Friedman, Michael Shirts

We present two different approaches for improving the utility of alchemical free energy calculation methods, each combining components from existing free energy methods in new and useful ways. Both approaches expand the range of applicability of the methods, and address different challenges that can arise in free energy calculations.

First, we present alchemical metadynamics, which allows explicitly biasing configurational collective variables simultaneously with alchemical free energy calculations, entirely within the metadynamics framework. Using test systems with varying complexity, we demonstrate that alchemical metadynamics can enable sampling of metastable states not necessarily accessible in standard free energy methods.

We also introduce the method of replica exchange of expanded ensemble (REXEE), where one which periodically exchanges coordinates of different replicas of EXE simulations that collectively span the range of interest. We show that this method computes free energies consistent with the estimates obtained in expanded ensemble and Hamiltonian replica exchange, both offering higher flexibility in how one allocates alchemical states between processors and lessening the need for tight coupling between processors. It also removes some of the issues that occur with expanded ensemble where state ranges can remain unsampled. We also provide examples of how the REXEE approach can allow multitopology simulations, such as chain growth or multi-ligand binding ensembles, without altering the underlying simulation code to support multiple topologies.

Predicting Gas-Phase Transport Properties of PFAS for Incineration Models Using Quantum Chemistry, Correlations, and Molecular Simulation

Tim Mallo, C. Claire Murphy, Hrishikesh Ram, Phil Westmoreland

Modeling the destruction of per- and polyfluorinated substances (PFAS) "Forever Chemicals" by incineration requires transport properties. In this work, we predict transport-property parameters via quantum chemistry and correlations. We are also beginning to examine and test these predictions of PFAS diffusivity by using molecular simulations of the relatively stiff, linear molecules. The result is a complete set of parameters for a separately developed kinetics model.

Current forms of Chemkin[1] and Cantera,[2] widely used codes for modeling gas-phase reactors, apply conventional kinetic-theory expressions based on Lennard-Jones potentials. The essential six species parameters besides mass are linearity (atom, linear, or nonlinear), Lennard-Jones diameter and well depth, dipole moment, dipole polarizability, and rotational relaxation collision number. Cantera adds Pitzer acentric factor, dispersion coefficient, and quadrupole polarizability.

The present work focuses on the first six parameters, where the dipole moment and polarizability parameters have been calculated using quantum chemical methods in Gaussian 16.[3] Structure optimization calculations are run at the B3LYP/6-31G(d,p) level, then later upgraded to M06-2X-D3(0)/def2-TZVP as selected in a separate study on predicting species thermochemistry.[4] Lennard-Jones parameters are calculated using Joback's group-additivity method. Rotational relaxation collision number represents the efficiency of a rotating molecule colliding with another molecule and losing rotational energy; based on complexity of the PFAS molecules, it is unity. These parameters are applied to different reacting flows and incinerator conditions, showing the potential impact of uncertainties.

The perfluoroalkyl chain is long and relatively linear, raising questions about appropriateness of using the Lennard-Jones spherical potential, so molecular simulations of the diffusivity are being carried out using LAMMPS.[5] That code offers two methods of generating diffusivities: (1) measuring the mean-squared displacement and relating its time rate of change to the diffusivity and (2) measuring the velocity auto-correlation function and relating its time integral to diffusivity. Preliminary results will be presented.

References

1.Chemkin Pro, Ansys. https://www.ansys.com/products/fluids/ansys-chemkin-pro

2. D.G. Goodwin, H.K. Moffat, I. Schoegl, R.L. Speth, B.W. Weber. Cantera: An object-oriented software toolkit for chemical kinetics, thermodynamics, and transport processes. https://www.cantera.org, 2023. Version 3.0.0. doi:10.5281/zenodo.8137090.

3. M.J. Frisch, G.W. Trucks, H.B. Schlegel, G.E. Scuseria, M.A. Robb, J.R. Cheeseman, G. Scalmani, V. Barone, G.A. Petersson, H. Nakatsuji, et al. Gaussian 16, Revision A.03. Gaussian, Inc., Wallingford CT, 2016.

4. H. Ram, T.P. Sadej, C.C. Murphy, T.J. Mallo, P.R. Westmoreland, 'Thermochemistry of Species in Gas-Phase Thermal Oxidation of C2 to C8 Perfluorinated Carboxylic Acids,' *J. Phys. Chem. A* **128**:7 (2024) 1313–1326. DOI: 10.1021/acs.jpca.3c06937.

5. A.P. Thompson et al., 'LAMMPS - a flexible simulation tool for particle-based materials modeling at the atomic, meso, and continuum scales,' *Computer Physics Comm.* 271 (2022) 108171. <https://www.lammps.org/>

Uncovering the Thermodynamic Driving Forces Underlying Sequence-Dependent Phase Separation of Proteins

Shiv Rekhi, Jeetain Mittal

Intrinsically disordered proteins can undergo sequence-dependent phase separation to form membraneless organelles (MLOs), which play a role in numerous cellular processes. A number of sequence features, termed as a molecular grammar, have been found to promote phase separation of disordered regions. Specifically, the presence of aromatic, usually Tyr, and charged residues, usually Arg, have been identified as primary drivers of phase separation. In this work, we design sequence variants of a model polypeptide targeting previously overlooked sequence features with the aim of interrogating the contributions of amino acids outside of the molecular grammar on sequence-dependent phase separation. We find that all but one of the sequences designed undergo phase separation, pointing to a model of all residues collectively contributing to the driving forces for phase separation rather than just a handful. To explain the thermodynamic forces underlying phase separation at the residue level, we perform alchemical free energy calculations to quantify the free energy of transfer of amino acid side chain analogs from the water-rich dilute phase to the protein rich dense phase. We uncover an interplay between protein and solvent-mediated interactions in determining the net transfer-free energies of all amino acids. Further, we dissect the contributions of protein, bulk water, and condensate water to the thermodynamic driving forces for phase separation. To summarize, in this work we uncover the contributions of residues previously overlooked in the molecular grammar of phase separation as well as provide a molecular picture of the forces underlying their contribution to the formation and stabilization of biomolecular condensates.

Dynamics and Interactions in Viral RNA Elements

Lev Levintov, Harish Vashisth

Ribonucleic acid (RNA) molecules play an essential role in the development of various diseases as well as in the replication and survival mechanisms of viruses. Therefore, viral RNA elements serve as a promising target for the development of novel therapeutic modalities to inhibit viral replication. Specifically, conserved and structured RNA elements in viral RNA genomes are inherently flexible and adopt complex three-dimensional folds that may provide transient binding pockets for therapeutic agents. Despite recent advancements in structure determination techniques which provide insights into the dynamics and interactions in viral RNAs, the characterization of all possible parameters that are required to describe the RNA dynamics at the atomic level of detail remains challenging. In this work, atomistic simulations were utilized to better understand structural and conformational effects of ligand binding on the dynamics of viral RNA elements. At first, ligand binding was observed to stabilize viral RNA structure due to the decreased flexibility in helical motifs which comprise the RNA. Then, the dissociation of a small inhibitor was identified to be coupled with a base flipping event which resulted from a sequence of complex events relating multiple nucleotides. The base flipping mechanism was further characterized using a refined reaction coordinate which describes the relative position of the flipping base with respect to the neighboring bases. The energetics and dynamics of methylation on an RNA hairpin capped by a methylated tetraloop motif was studied, revealing enhanced flexibility upon methylation. Lastly, a network of hydrogen bonding and salt bridging interactions was determined to facilitate the recognition mechanism between a helical peptide with inhibitory properties and a viral RNA element.

Benchmarking Machine Learning Potentials to Reproduce Thermodynamic and Transport Properties of Bulk Liquids.

Eliseo Marin Rimoldi, Edward Maginn

Classical molecular dynamics (MD) and Monte Carlo (MC) methods rely on force fields to describe the underlying potential energy that governs molecular interactions. A recurrent strategy to parametrize force fields for organic liquids relies on the availability of thermodynamic and transport experimental measurements, such as density or viscosity, which might not be available for the system under study at the conditions of interest. An alternative route to obtain such properties is to conduct ab initio molecular dynamics (AIMD) simulations, which do not require pre-existing experimental measurements. Despite this clear advantage, AIMD is prohibitively expensive, severely limiting the accessible length and time scales required to compute the physical properties of interest.

Recently, neural networks have raised a tremendous amount of interest due to their ability to reproduce complex and multi-dimensional potential energy functions with high accuracy. They have been proposed as surrogates to costly AIMD simulations, as they faithfully reproduce the underlying quantum mechanical energies and forces with less computational demands. Thus, longer time and length scales can be readily available. We explore the ability of neural network force fields that use local descriptors to reproduce transport and thermodynamic properties of several organic liquids. We trained machine learning force fields using energies and forces from reference AIMD simulations. Additionally, we explore an active learning approach to train machine learning force fields more efficiently. We show that the trained neural networks reproduce the transport and thermodynamic properties of interest relative to the reference AIMD calculations.

Efficient Bayesian Machine Learning Interatomic Potentials for Materials Science Research

Hao Deng, Bin Liu

In this work, we report a unified sparse Gaussian process (SGP) machine learning potential for the phase stability predictions of boron allotropes (i.e., α-B, β-B, and γ-B) based on an active, on-the-fly learning training process. The employed SGP model enables a significant reduction of the training data size. Furthermore, by incorporating an active learning scheme, the most relevant structures are produced on the fly.

Elemental boron has attracted much attention due to its superior and versatile properties. Among the numerous allotropes of boron, α-B (rhombohedral, B_{12}), β-B (rhombohedral, B_{105}), and γ-B (orthorhombic, B_{28}) all consist of B_{12} icosahedra as the fundamental building block. The trained SGP yielded good agreement with DFT on predictions of structural, thermodynamics, and vibrational properties. Insights were gained from the generated P-T diagram and free energy composition analyses. The β phase dominates the area corresponding to high temperatures and low pressures, while γ-B occupies the high-pressure region. A notable stabilization due to the B13 vacancy was observed, which validates the extrapolation ability of the SGP potential trained from the defect-free training set. We anticipate this computational approach will aid the understanding of complex crystals in future studies, such as multi-element materials with vacancies, which are important for energy conversion applications.

Digital Alchemy for the Inverse Design of Patchy Particles

Timothy Moore, Luis Y. Rivera-Rivera, Sharon C. Glotzer

Patchy particles are an exciting development for the fields of colloidal and nanomaterials. The ability to tailor interparticle interactions offers the promise of precise control of the self-assembly of patchy particles into structures with desirable properties. Given the many possible "degrees of anisotropy," the design space of patchy particles is enormous, prohibiting the direct exploration of this space and making inverse design methods a crucial tool for the field. Of the many inverse design methods developed to date, digital alchemy is particularly attractive for patchy particles. Digital alchemy is an extended ensemble method that incorporates particle properties into the thermodynamic ensemble, allowing inverse design via statistical-mechanically-rigorous, on-the-fly updates to interparticle pair potentials. In this work, we extend the digital alchemy framework for the inverse design of patchy particles, and design triblock Janus particles to self-assemble target crystal structures. We provide examples of symmetric triblock Janus particles in 2- and 3-dimensions that self-assemble the open kagome and pyrochlore lattices, respectively. To highlight the generality of our method, we also present the inverse design of asymmetric triblock Janus particles to self-assemble a snub square lattice, whose coordination has a lower symmetry than the kagome and pyrochlore lattices, and therefore requires patches of differing sizes. Surprisingly, particles designed based upon geometric considerations are unable to self-assemble into the snub square lattice, whereas particles designed via our digital alchemy framework do, highlighting the ability of the method to find nontrivial solutions to the design problem. This work shows how digital alchemy can be used to design arbitrary patchy interactions to self-assemble into desired crystal structures, and can be built upon to design patchy particles of arbitrary shape and patch configuration.

Extracting Structures and Transferrable Pair Potentials from Experimental Neutron Scattering Results

Michael Hoepfner, Brennon L. Shanks, Abdur R. Shazed, Harry W. Sullivan

Experiments are samples drawn from a perfect simulation, containing all the quantum mechanical, thermodynamic, and complex interactions of the real world. In principle, experimental techniques that sample these phenomena directly, such as X-ray and neutron scattering, could be analyzed to learn these fundamental physical interactions and hence to guide molecular simulation development. We have recently demonstrated early progress on this challenge using a new technique, Structure Optimized Potential Refinement, where pair interactions extracted from neutron scattering results have been shown to reproduce the local fluid structure observed experimentally and demonstrate accurate prediction of ensemble thermodynamic and transport properties at states far from the optimization conditions. Consequently, we anticipate that detailed insight into local atomic forces can be obtained from experimental scattering measurements, providing a novel pathway for force field optimization and insight into atomic physics. In this poster, we will provide an overview of progress to date and discuss the remaining challenges associated with strategies to extract accurate structures and atomic forces from experimental results.

Solvent Induced Reconfiguration of Nanoparticle Superlattices

Tobias Dwyer, Yaxu Zhong, Timothy C. Moore, Jun Chen, Yi Wang, Alex Butrum-Griffith, Xingchen Ye, Sharon Glotzer

The self-assembly of nanoparticle superlattices has been a challenge widely achieved in recent years. However, imaging the entire assembly process on a per-particle level, let alone predicting the entire pathway of assembly is a grand challenge that the molecular simulation community has yet to achieve. If we want to design the process of assembly and reconfigurability between different states, it is essential that we can predict the entire assembly process as well as the final structure on a per particle level. In this work, we take the first major step in this challenge by tuning the self-assembly of gold nanocubes under TEM illumination by simply changing the solvent in which the nanoparticles self-assemble. Under Illumination, we hypothesize that the nanoparticles become charged by the TEM beam, while the solvent undergoes radiolysis which screens the electrostatic repulsion between nanoparticles while attraction between nanoparticles is induced via Van der Waals (VdW) forces. By building a simulation model that captures VdW attraction and electrostatic repulsion, we accurately predict the phases that assemble in experiment and their assembly pathway. Furthermore, through dimensional analysis, we show that the self-assembly process can only be explained via a change in the screening length of the solvent and not via a change in the magnitude of repulsion between the nanoparticles. Additionally, we use our model of electrostatic repulsion and VdW attraction to predict the boundary between orientationally ordered and disordered superlattices, allowing us to tune the assembly of rotator crystals. Finally, we leverage our understanding of different solvents to in-situ reconfigure our lattices and compare the mechanism of reconfiguration to simulations. Our combined simulations, dimensional analysis, and experimental verification thereby elucidate the mechanism of self-assembly for these reconfigurable nanoparticle superlattices. This work is the first step in designing reconfigurable nanoparticle matter whose pathways of reconfiguration and final structure can be observed in-situ and predicted by simulation and theory.

Machine-Learned Committor Functions for Reactive Molecular Dynamics

Jacob Gissinger

Reactive molecular dynamics (MD) is a powerful tool for atomistic-scale modeling of a diverse range of chemical processes. However, scaling these simulations to large systems and long times scales remains a challenge because of the complexity of the potential energy function required. The author previously developed a heuristic approach, called REACTER, that incorporates reactivity in MD simulations in a less general but much more computationally efficient manner. REACTER uses standard, fixed valence force fields as the underlying potential energy surface for describing all interatomic interactions but adds a procedure for enforcing user-defined reactions that occur when certain geometric constraints on relative atomic positions are satisfied. Further, these bonding changes can be accepted or rejected with a probability related to the local thermal energy. This work seeks to generalize this approach by replacing the set of user-defined geometric constraints and energetic criteria with a committor function that specifies the probability of a reaction occurring on the basis of the local atomic configuration. The committor function is a useful mathematical tool for modeling rare events but, unfortunately, is very difficult to compute for realistic systems in a general way. This work describes a method for approximating the committor function using a machine learning approach, specifically a deep neural network trained with data from reactive MD and DFT-based dynamics simulations. This network is coupled to the existing REACTER protocol, as implemented in the LAMMPS MD package, and used to make on-the-fly predictions of reaction probabilities without the more extensive user input previously required. The new method is demonstrated using the polymerization of polystyrene as a case study. Although very dependent on the quality and quantity of training data, machine-learned committor functions show promise as a method for incorporating reaction probability from higher level calculations into highly scalable MD simulations.

Evaluating the Performance of Semi-Empirical Electronic Structure Methods to Model Phonon Transport in Metal–Organic Frameworks

Masoumeh Mahmoudi Gahrouei, Nikiphoros Vlastos, Laura de Sousa Oliveira

Metal–organic frameworks (MOFs) are two or three-dimensional ultrahigh porosity materials that consist of metal nodes connected by organic linkers. MOFs have the potential to be used as thermoelectric materials. Thermoelectric materials *use temperature differences to generate electrical energy*. The main motivation for this project is to further our knowledge of thermoelectric properties in MOFs and find which available self-consistent-charge density functional tight-binding (SCC-DFTB) method can best predict (at least trends in) the phononic part of the thermoelectric properties of MOFs at a lower computational cost than standard density functional theory (DFT). To discover MOFs with desired TE properties, it is important to investigate stability and thermal properties of MOFs that can be understood from phonon properties. We compared phonon transport properties obtained with SCC-DFTB/3ob and SCC-DFTB/mio (based on available Slater-Koster files), GFN1-xTB and GFN2-xTB against PBE-DFT for $Zn_3C_6O_6$, Zn-NH-MOF, $Ni_3(HITP)_2$ and $Cd_3C_6O_6$ MOFs for the monolayer and stacked AA, serrated and AB structures. The phonon dispersion results using GFN2-xTB and DFTB-3ob/mio resemble the DFT-PBE results, and we therefore predict that the lattice thermal conductivities from GFN2-xTB and, to a lesser extent, DFTB-3ob/mio are closer to the DFT-PBE results. GFN2-xTB is expected to outperform other methods in predicting the lattice thermal conductivity, given that the optimized geometries (i.e., lattice parameters, bond lengths, angles, etc.) using GFN2-xTB compare favorably to the optimized geometries using DFT-PBE. Based on the phonon dispersion results, the serrated Zn-NH-MOF and $Zn_3C_6O_6$ are the most stable structures. Curiously, GFN2-xTB did not converge for many unstable geometries (i.e., those with negative phonon modes) and was found to converge for all but one of the stable geometries. We also calculated the phononic contribution to the thermal conductivity using SCC-DFTB/3ob(mio) and GFN-xTB. While $Zn_3C_6O_6$, Zn-NH-MOF and $Cd_3C_6O_6$ MOFs are predicted to have a higher power factor than the $Ni_3(HITP)_2$ MOF (one of the highest performing synthesized thermoelectric MOFs), thermal transport is substantially lower for Zn-NH-MOF, leading to the highest ZT among the group.

Simulating the Flow of Polymer Solutions in Sinusoidal Channels Using Dissipative Particle Dynamics

Clara De La Torre Garcia, Tzortzis Koulaxizis, Antonia Statt, Michael Howard

The flow of dilute polymer solutions through contraction–expansion channels can exhibit a surprising transition to turbulence at low Reynolds numbers. This "elastic turbulence" has been studied experimentally and using analytical & computational fluid-mechanics treatments, but the molecularly detailed picture of this phenomenon is still incomplete. Here, we use dissipative particle dynamics (DPD) to simulate the flow of dilute solutions of linear polymers through sinusoidally varying microchannels. We first validate our DPD approach by simulating the flow of a Newtonian solvent and comparing our simulations to an analytically predicted flow field. We then characterize the polymer conformations and flow of non-Newtonian polymer solutions in the same geometry, aiming to identify signatures of turbulence. Our simulations have important implications for understanding processes involving flow through porous media, such as enhanced oil recovery and environmental remediation.

Lessons in Modeling Electron Transport of Two-Dimensional Covalent– and Metal–Organic Frameworks for Semiconducting Applications: A Benchmarking Study

Masoumeh Mahmoudi Gahrouei, Alathea Davies, Nikiphoros Vlastos, Laura de Sousa Oliveira

As devices become increasingly more efficient, materials functionalization and tunability take a front seat to efficiency in the quest for novel semiconductors. Covalent– and metal–organic frameworks (C/MOFs) can be functionalized and offer high porosities and surface areas, lending themselves to multifunctional materials applications, especially if their electronic properties can (also) be tuned. Recent computational work by our group has required benchmarking the electronic properties of COFs and MOFs with semi-empirical electronic structure methods (SCC-DFTB and GFN-xTB) against first principles approaches (DFT-PBE and HSE06). The work presented here distills some of the computational best practices and lessons learned in our search for COFs and MOFs suitable for semiconducting applications.

Efficient Sampling of Protein Conformational Dynamics with Anm-Guided Ptmetad

Betul Uralcan

Exploring the complex conformational dynamics of proteins is crucial for understanding essential biological processes. Molecular simulations offer insights into these dynamics, but they are often limited by timescale constraints. Enhanced sampling techniques, such as parallel-tempering metadynamics (PTMetaD), address this limitation by efficiently sampling the free energy landscape. However, PTMetaD's effectiveness is hindered by the choice of collective variables (CVs). In this study, we introduce a novel framework for designing efficient CVs for PTMetaD, aiming to minimize hysteresis effects. Our approach utilizes slow modes derived from α-carbon-based anisotropic network models (ANM) as CVs, effectively guiding the system along the highest collective modes of motion. By only using the intrinsic dynamic modes of motion that are accessible to the protein, our framework significantly enhances convergence of the free energy landscape compared to conventional PTMetaD based on CVs. Through simulations of proteins, including Adenylate Kinase (AdK) and a G protein–coupled receptor (GPCR), we demonstrate the efficiency and reliability of our method in characterizing large-scale conformational changes and associated free energy landscapes.

Estimation and Uncertainty Quantification of Experimental Radial Distribution Functions Using Non-Stationary Spectral Gaussian Processes

Harry Sullivan, Brennon Shanks, Michael Hoepfner

Across interdisciplinary boundaries spanning materials science, chemistry, and biophysics, understanding the radial distribution function provides valuable insights into various material properties through Kirkwood-Buff integration and is used as a vital benchmark for molecular simulation. Unfortunately, directly measuring the radial distribution function for atomistic systems is not possible. Instead, measurements are conducted on its counterpart, the structure factor, which are related through a radial Fourier transform. While this transform is theoretically sound, practical implementations encounter obstacles. Firstly, the finite size of individual neutron detectors limits measurements to discretized sets of momentum transfers. Secondly, limited angular detector coverage confines measurements to a specific range. Lastly, measurement uncertainty, including neutron count and momentum transfer position (e.g., time-of-flight uncertainty), introduces noise that can distort the underlying distribution. These distortions can cause force field optimization algorithms such as structure optimized potential refinement, iterative Boltzmann inversion, and relative entropy minimization to overfit, resulting in low transferability or inaccurate thermodynamic predictions. To enable comparisons between molecular simulations and experimentally-determined measurements of fluid local structure, a method capable of addressing all these errors concurrently is essential. Here, we are investigating the impact of imperfect experimental data on real-space structural conclusions by employing Non-Stationary Spectral Gaussian Processes. These machine learning strategies offer a distinct advantage over existing techniques, being fully nonparametric, providing built-in uncertainty quantification, and resting on a well-understood mathematical foundation.

Exploring the State-Dependence of Classical Pair Potentials with Neutron Scattering

Brennon Shanks, Harry Sullivan, Michael Hoepfner

Pairwise additive interatomic forces are known to change with thermodynamic state (temperature, pressure, density), meaning that a valid pair potential model that gives good results for one set of thermodynamic states is not guaranteed to accurately model thermophysical properties at others. An important unanswered question is precisely how and why these interatomic forces change with thermodynamic variables, and if so, can we measure these effects experimentally? One direct way of exploring state-dependent behavior is to investigate structural correlation functions derived from neutron scattering experiments. Structural correlation functions, which heavily influence our understanding of self-assembly and thermodynamic properties of liquids, arise from nuanced interatomic interactions that can be inferred using structural inversion techniques. Here we study the thermodynamic dependence of classical forces in liquids using a neutron scattering dataset of fluid krypton. Parametric (n-6) Mie potential models learned using accelerated Bayesian inference and nonparametric pair potentials estimated using structure optimized potential refinement (SOPR) are compared.

Flexicles: A Computational Model to Study the Individual and Collective Behavior of Active Cell-like Microrobots

Philipp Schönhöfer, Sophie Y. Lee, Sharon C. Glotzer

Although the boundary between the functionality of synthetic active microparticles and of their biological counterparts become increasingly blurred, we still lack the fundamental understanding to recreate the key facets of autonomous behavior exhibited by microorganisms or macroscopic robots. In this study, we propose a model for a three-dimensional self-driven, deformable cellular robot composed of self-propelled particles confined to a flexible membrane - a superstructure we call a "*flexicle*". Using molecular dynamics simulations, we investigate the behavior and reaction mechanisms of individual flexicles within complex environments or within a collective. We show that flexicles can adopt to their surroundings by exhibiting intricate shape changes. This coupling between morphological adaptability and particle propullsion gives rise to a diverse set of behaviors from motility-induced phase separation phenomena and the spontaneous flow of flexicles in multi-flexicle systems to curvature sensitive dynamics and overcoming topographcal barriers of single flexicles. Our findings establish a foundation for controlling the migration of cell-like active particles and for developing strategies for achieving other autonomous robotic behaviors.

Sopr Pair Potentials for Aromatic Compounds: Revisiting Assumptions in Force Field Database Formulation and Structural Representation

Abdur Rahman Shazed, Harry W. Sullivan, Michael Hoepfner

Aromatic compounds, characterized by delocalized π-electron systems, are ubiquitous in organic chemistry, materials science, and pharmaceuticals. Their unique electronic structure confers diverse properties, making them essential in various applications. In this study, we delve into the intricacies of force field databases for aromatic compounds, focusing on the application of Structure Optimized Potential Refinement (SOPR) pair potentials. We challenge the common assumption of treating similar atoms from different compounds as having identical model parameters, revealing that this assumption may not hold true when aiming to accurately match structural features. Our investigation further aims to uncover the necessity and origin of the multi-well oscillating nature of SOPR potentials that we observed when the SOPR algorithm was applied for benzene, methane, and water for experimental structural representation. Though the multi-well SOPR potential demonstrated excellent transferability in predicting both thermodynamic and transport properties, to enhance applicability of the SOPR potential it is prudent to justify the necessity and origin of the multi-well nature of the potential. Given the intricate nature of the challenge, which necessitates the simultaneous optimization of SOPR algorithms across multiple aromatic systems, there arises a potential need for a machine learning approach. Specifically, employing active learning methods could offer a promising strategy to mitigate computational costs.

Effects of Single Atom Modifications on Protein-Ligand Interactions Studied Using Markov State Modeling

Sumit Sharma, Ramin Mehrani, Jagannath Mondal

In protein-ligand interaction studies, it is often observed that small changes in the structure of ligands can have a large influence on their binding affinities on proteins. For example, recent experimental studies of ligand binding on glycogen synthase kinase -3 (GSK-3) have shown that changing the position of nitrogen on the pyridine ring of ligands has a dramatic effect on their inhibition efficiency. In this work, we have implemented Markov State Modeling in molecular dynamics simulations to study if one can discern the influence of these small changes in the ligand structure on their binding affinity on GSK-3 protein. We find that our modeling is able to capture the effect of single atom modifications on the binding energy as well as provide insights into the underlying reasons for this observation.

Hiking up the Energy Landscape: Using Molecular Dynamics and Forward Flux Sampling to Probe Ion Transport through Nanoporous Membranes

Omar Khalifa, Brian Shoemaker, Amir Haji-Akbari

Membrane-based separation technologies, with applications ranging from desalination to resource recovery, are of increasing interest. These processes may involve ion transport through nanopores, where classical theories often fall short in explaining structure-selectivity relationships. Achieving high spatiotemporal resolution in experiments is still under quest, leaving considerable gaps in our understanding of how structure impacts selectivity. Availability of computational power and advances in simulation techniques, particularly advanced sampling techniques, provide valuable tools to address this gap. In this work, we investigate ion and water transport across nanoporous membranes and their sensitivity to structural factors, including pore proximity and charges.

Using path sampling simulations, namely forward flux sampling (FFS), we model graphitic membranes with sub-nanometer pores that are decorated with permanent dipoles, where we study the effect of pore proximity on permeability and selectivity at a given pore density. We reveal that there are pore-pore correlations that arise with pore proximity, affecting the passage of ions and solvent. Particularly, these pore-pore correlations detrimentally affect the passage of the counter-ions and the polar solvent (water), while enhancing the passage of the co-ions. Overall, both water permeability and salt rejection are reduced with proximity. This finding deviates from theories such as the access resistance model. We propose a new simple phenomenological model that considers the pre-pore ion concentration and the forces exerted on the traversing ion through the dipole-decorated pore to predict the relative extent of the pore-pore correlations on ions selectivity of multi-pore membranes. We then use this model to screen multiple pore arrangements that exhibit the least pore-pore correlations. In this work, we also investigate the effect of the pore's permanent dipole strength in terms of partial charges and how does that affect ion selectivity and water permeability. Preliminary results show that the counter-ion traverses faster through pores with stronger dipoles. It is also expected for co-ions to traverse slower in stronger charged dipoles due to the lesser electrostatic pull. We are conducting more simulations to predict the total salt rejection of water permeability in stronger charged dipoles systems and compared it to the weaker charged dipoles systems. Such findings of ion transport through membranes are subject to change due to the presence of strong finite size effects. Mitigating these effects is necessary to affirm the validity and qualitative accuracy of the simulations. In this study, we also investigate the extent of these finite-size effects that appear due to extraneous electrostatics imposed by the periodicity of the simulations box. We developed a physics-based model to correct for them and obtain corrected free energy profiles, which allows us to predict more accurate passage times and selectivities. This work aids in informing the design of more efficient membranes, and understanding nanoscale mechanisms of separations, including biological processes.

Monte Carlo Sampling of Transition-State Structures for Zeolite-Catalyzed Reactions

Shubham Malviya, Peng Bai

Nanoporous materials such as zeolites are often described as robust, enzyme-like catalysts. They exhibit exquisite molecular shape selectivity due to the interactions offered by the extended, heterogeneous reaction environments. This characteristic, however, presents significant challenges to the predictive modeling of reactions occurring insides these materials, especially for large, articulated reactants. This work presents a computational procedure that couples zero-Kelvin quantum-chemical methods and forcefield-based Monte Carlo sampling to enable a consistent calculation of local activation energies for bulky reactants. Using this approach, we studied the protolytic cracking of n-butane in the TON, MFI, LTA, and FAU zeolites. It is found that in the MFI zeolite while the tighter zig-zag and straight channels have lower transition-state (TS) energies, the more spacious intersection region potentially supports a much larger number of TS configurations, suggesting a potentially higher TS entropy. To compare across different zeolites, we attempted to compute both site-specific and ensemble-averaged intrinsic activation barriers. In one approach, the reactant-state energies were calculated by restricting them to those configurations that are closest to the active site. We found that subtracting such "local" reactant energies does not remove the local confinement effects but rather exacerbates the difference, with site-specific intrinsic barriers ranging from 184 to 219 kJ/mol. We argued that for kinetics-limited reactions, a more appropriate reactant state should be globally defined for each zeolite. Using this approach and further examining two assumptions about the acid site distribution, either stochastic or according to their formation energies, ensemble-averaged intrinsic barriers were computed. These site-averaged barriers were much more similar across the zeolites, with three zeolites (MFI, FAU, and LTA) all having intrinsic barriers of ~220 kJ/mol and the TON zeolite around 200 kJ/mol.

A Computational Perspective on Polyester Adsorption Behavior to Alumina at the Catalytic Depolymerization Interface

Ziyue Dong, Aidan Wegner, Michael F. Toney, Kayla G. Sprenger

Upcycling end-of-life plastics into alternative chemicals is critical to envisioning a waste-free and circular plastic economy. Hydrogenolysis, using supported metal catalysts, has shown promise as a viable catalytic pathway for transforming plastic wastes into new value-added chemicals in an economical manner. However, the complex chemical makeup of mixed plastic waste streams poses a major challenge to realizing this opportunity. Understanding the key molecular interactions that take place at the inorganic-organic depolymerization interface and directly affect catalytic performance, from a computational perspective, is thus paramount to improving the design of new catalysts for plastics hydrogenolysis.

We hypothesized that interactions between the polymers and the catalytic surface would be governed by the presence of the metal oxide support. As such, we performed physics-based, molecular dynamics simulations on the polymer-support interfacial system to elucidate these key interactions. Here, we report the interfacial molecular layering, orientational distribution, and free energies of interaction, of polyester polylactic acid and polyethylene terephthalate, as well as polyolefin polyethylene, at the alumina metal oxide support surface. Our simulation results identified variable effects of polymer chain length and functional groups on the interfacial molecular behavior of the polymers, which will be validated using X-Ray reflectivity experiments on analogous *in vitro* model systems. In this investigation, we applied state-of-the-art enhanced sampling techniques well-tempered metadynamics and parallel tempering metadynamics in the well-tempered ensemble. These methods rendered highly accurate descriptions of polymer-support interactions and near-surface polymer configurations over timescales not previously achievable using classical simulations. Molecular insights gained from this work can be leveraged to improve future catalyst designs and propose catalytic mechanisms that directly affect reactivity and selectivity. The general simulation workflow derived from this work can also be applied to elucidate interfacial phenomena at other inorganic-organic interfaces, such as that of lithium-ion battery electrodes and of organically-functionalized colloidal nanocrystals.

Molecular Dynamic Simulations of a Disordered Drug-Loaded Block Copolymer Micelle in Aqueous Solutions

Negin Razavilar

In previous investigations, all-atom molecular dynamics (MD) simulations of a single hydrophobic drug molecule in a series of poly (ethylene oxide-b-caprolactone) (PEO-b-PCL) pseudo-micelles were conducted to understand drug-polymer interactions and drug diffusion. While these simulations provided valuable information about the hydrogen-bonding interactions, it was not possible to capture the full effect of the interactions on the drug diffusion dynamics. Atomistic MD simulations of drug diffusion from stable drug-loaded micelles are costly and impractical due to the very long timescales associated with drug release from such systems. To address this computational challenge, our present study initiates an all-atom MD simulation from a disordered structure of a full Cucurbitacin B (CuB)-loaded PEO-b-PCL block copolymer micelle in aqueous solutions. Our findings reveal that the ensuing drug diffusion out of the micelle is initially facilitated by the formation of the water-PCL bonds and the subsequent disruption of water-PEO bonds, followed by the formation of water-CuB bonds.The dynamics of CuB and water exhibit non-linear sub-diffusive mean-squared displacements, attributed to molecular crowding within the micelle and hydrogen bonding interactions between water/CuB molecules and polymer chains. Significantly, our study indicates that the hydrogen bonding and diffusion dynamics observed in the pseudo micelle are not representative of those occurring in the full micelle environment. This computational approach is anticipated to provide detailed molecular-level insights that contribute to a better understanding of in-vitro drug release data from nano-sized polymer micelles.

Multiscale Modeling of Transport Properties of KOH (aq) with Machine Learning

Vladimir Lagerweij, Parsa Habibi, Sana Bougueroua, Poulumi Dey, Thijs J.H. Vlugt, Marie-Pierre Gaigeot, Othonas A. Moultos

Aqueous potassium hydroxide exhibits an exceptionally high electrical conductivity, which is crucial for various industries reliant on efficient electrolytes. This increased conductivity arises from the Grotthuss mechanism, where hydroxide ions exchange protons with water molecules. Applications of KOH(aq) range from alkaline water electrolysis to electrochemical CO_2 reduction. Even the most state of the art classical force fields for molecular dynamics fails to predict OH^- diffusion correctly, as it cannot capture the Grotthuss mechanism.[1] Ab initio molecular dynamics (AIMD) captures the proton transfer, but is limited by short timescales and small system sizes. Machine learning force fields (MLFF) have emerged as a promising solution, offering longer simulation times and near-AIMD accuracy. In this work, an MLFF capable of accurately capturing OH^- transport behavior is constructed. Structural properties of AIMD and MLMD simulations are compared to ensure accuracy of the MLFF. Graph theory is applied to generate 2D representations of the molecular structures [2], which are used to analyze the hydration the OH^- during proton hopping events. This method distinguishes the two hydration modes, the reactive and inactive modes, which supports the qualitative findings of Tuckerman et al.[3] MLMD simulations at experimental density, 345 K, and with 1 mol KOH per kg of water, yield a self-diffusion coefficient for OH^- ions of $(6\pm2)\times10^{-9}$ m^2/s, closely matching the experimentally-found value 5.27×10^{-9} m^2/s. The computed electric conductivity of (29±6) S/m overlaps with the experimentally determined conductivity of 26 S/m as well. Our work not only advances the understanding of OH^- transport, but it also shows the effectiveness of ML in molecular simulations. It demonstrates the potential of MLFFs to bridge the gap between classical MD and AIMD simulations, and highlighting the utility of graph theory in understanding molecular structures, while offering valuable insights for industries reliant on efficient electrolytes as well.

[1] Habibi, P.; Rahbari, A.; Blazquez, S.; Vega, C.; Dey, P.; Vlugt, T. J. H.; Moultos, O. A., A New Forcefield for OH^- for Computing Thermodynamic and Transport Properties of H_2 and O_2 in Aqueous NaOH and KOH Solutions. *J. Phys. Chem. B* **2022**, *126*, 9376–9387.
[2] Bougueroua, S.; Spezia, R.; Pezzotti, S.; Vial, S.; Quessette, F.; Barth, D.; Gaigeot, M.-P., Graph Theory for Automatic Structural Recognition in Molecular Dynamics Simulations. *The Journal of Chemical Physics* **2018**, *149*, 184102
[3] Tuckerman, M. E.; Marx, D.; Parrinello, M. The Nature and Transport Mechanism of Hydrated Hydroxide Ions in Aqueous Solution. *Nature* **2002**, *417*, 925–929

Understanding the Surprising Ionic Conductivity Maximum in $Zn(TFSI)_2$ Water/Acetonitrile Mixture Electrolytes

Yong Zhang, Emily Carino, Nathan Hahn, Nigel Becknell, Julian Mars, Kee Sung Han, Karl Mueller, Michael Toney, Edward Maginn, Sanja Tepavcevic

Aqueous electrolytes composed of 0.1 M Zinc bis(trifluoromethylsulfonyl)imide ($Zn(TFSI)_2$) and acetonitrile (ACN) were studied using combined experimental and simulation techniques. The electrolyte was found to be electrochemically stable when the ACN V% is higher than 74.4. In addition, it was found that the ionic conductivity of the mixed solvent electrolytes changes as a function of ACN composition and a maximum was observed at 91.7 V% of ACN although the salt concentration is the same. This behavior was qualitatively reproduced by molecular dynamics (MD) simulations. Detailed analyses based on experiments and MD simulations show that, at high ACN composition, the water network existing in the high water composition solutions breaks. As a result, the screening effect of the solvent weakens and the correlation among ions increases, which causes the decrease in ionic conductivity at high ACN V%. This study provides a fundamental understanding of this complex mixed solvent electrolyte system.

Modeling Nucleation and Other Rare Events with Deep Learning Augmented Molecular Simulations

Vanessa Meraz, Ziyue Zou, Pratyush Tiwary

Nucleation is pertinent mechanism to crystal formation seen across a spectrum of scientific problems, and presents challenges that span a large temporal and spatial range. Molecular dynamics (MD) simulations provide the means to explore this domain, although MD is not without shortcomings, namely timescale limitations. This lends to the necessity in finding suitable descriptors for the crystallization process. In our case, we learn a low-dimensional reaction coordinates to describe the phase transformation for crystallization. In this work, we simulate the crystallization of supersaturated colloid suspensions using biased metadynamics with a reaction coordinate expressed as a linear combination of input features. The reaction coordinate is learnt using the State Predictive Information Bottleneck framework. In being computationally cost-effective, we demonstrate the sufficiency of using a one-dimensional reaction coordinate to capture relevant slow modes in a systems' dynamics. Building off of these results, we'll likely further discuss the application of expressive neural networks in more complex settings.

Hindered Transport of Water in Defective or Functionalised Carbon Nanopores.

Quang K. Loi, Debra Bernhardt

The transport properties of adsorbed molecules under nanoconfinement are important to the tailored design of nanoporous materials for various technological applications. The internal mass transfer is strongly dependent on the nature of the molecular interaction as well as the complex topology and interconnectivity of the porous network. Water is known to be hydrophobic on highly graphitised surfaces, but it displays an unusual trend in diffusion coefficients on different types of activated carbons. In this study, we performed equilibrium molecular dynamics simulations of water adsorption in slit-shaped pores composed of idealised, defective, or functionalised graphene layers to investigate the highly correlated nature of diffusion. We demonstrated the transition in the dynamics of adsorbed water as a function of pore sizes and the heterogeneity of the graphene surfaces. As the pores become non-idealised, water tend to cluster and the molecular transport ranges from mobile clusters to individual hopping of water molecules between neighbouring clusters surrounding a strong site, be it a functional group or a defect.

Investigating How HIV-1 Antiretrovirals Differentially Behave As Inhibitors and Substrates of P-Glycoprotein Via Dynamic Binding

Daisy Fuchs, Lauren Serio, Kayla Sprenger

HIV can penetrate and establish an infection in the brain within as little as a week after the initial exposure. The presence of HIV in the brain commonly results in neuronal death and a range of neurocognitive dysfunctions collectively called HIV-Associated Neurocognitive Disorder, or HAND. Combination antiretroviral drug therapies have been used since the 1990's to treat HIV infection by suppressing the viral load to undetectable levels. However, the ability of these small molecules to penetrate and treat HIV infection in the brain is severely hindered by transport barriers posed by the blood-brain barrier (BBB). The efflux protein P-glycoprotein (P-gp) is upregulated in the BBB and prevents therapeutic drug concentrations from being achieved within the brain. Despite being one of the most widely studied efflux proteins, there is no FDA-approved P-gp inhibitor, in part due to the severe side effects of complete inhibition of P-gp. HIV-1 protease inhibitors (PIs), a class of antiretrovirals, have been shown to behave as partial P-gp inhibitors, offering potential to modulate P-gp dynamics without compromising the essential protective functions of P-gp.

Herein, we have described how molecular dynamics simulations can be used to investigate the behavior of PIs as both substrates and inhibitors of P-gp. Our work has found that both dynamics and free energy of binding are crucial in understanding how different PIs interact with P-gp, especially in the binding pocket. Part of this work has explored how the binding pocket changes over the course of efflux, expanding the ideas of how substrate biding can occur and has demonstrated that PIs can be competitive inhibitors of P-gp. We have also taken into consideration that P-gp is an inherently dynamic protein and have studied how different starting conformations of P-gp impact the interactions between protein and small molecule inhibitors/substrates. A better understanding of these atomistic interactions allows for more insight on using PIs as P-gp inhibitors and future design of P-gp inhibitors.

Mapping the Design Space of Metal-Organic Frameworks As a Platform for Entangled Photon Pair Generation

Wilson Raney, Yamil Colón

Metal-organic frameworks (MOFs), crystalline networks of metal nodes and organic linkers, have recently been shown to display nonlinear optical (NLO) capabilities for important quantum processes like spontaneous parametric down-conversion (SPDC) and second-harmonic generation (SHG). Experimental and *ab initio* computational efforts to characterize the NLO properties of MOFs are still in their infancy. The design space of MOFs is substantial considering all possible combinations of nodes, linkers, topologies, and functional groups that can compose the crystal. A framework to characterize the entire design space of MOFs with NLO capabilities is needed to unlock the potential of MOFs to serve as materials for entangled photon generation. We applied *ab initio* techniques to understand the relationship between MOF chemistry and entangled photon generation by simulating the MOF MIRO101 with several functional groups. We calculated dielectric and second-order susceptibility tensors in CRYSTAL to obtain the ordinary and extraordinary refractive indices (n_0 and n_e), Sellmeier coefficients, phase matching angle, the effective nonlinear coefficient (d_{eff}), and the photon pair correlation time (τ_c). We further calculated the polarizability tensors of individual MOF linkers to correlate intrinsic electronic properties of the material with NLO response of the entire crystal. This work lays the foundation for a wider exploration of MOFs as candidate quantum optics systems by identifying top-performing materials from large libraries of theoretical structures.

Accurate Representation of Hydrogen in Metals By Machine-Learning Enhanced Modelling of Nuclear Quantum Effects

Kai Sellschopp, Thomas Klassen, Paul Jerabek, Claudio Pistidda

In a sustainable economy built on renewable energy, hydrogen plays a key role for storing energy and replacing fossil fuels. An efficient way to store hydrogen is to keep it in the solid state by binding it chemically in a metal hydride, which is particularly useful for seasonal energy storage or for applications where safety is a concern. Hydrogen is known to show nuclear quantum effects (NQE), such as high delocalization and tunneling through energy barriers, even at elevated temperatures. Additionally, the bonding situation in many metal hydrides gives rise to an anharmonic potential energy surface (PES) in the vicinity of typical hydrogen positions. Technically, an accurate representation of Hydrogen in metals therefore needs to take anharmonic NQE into account, but these have been mostly neglected in computational studies so far due to the high cost of the associated path-integral molecular dynamics (PIMD) calculations.

In this work, a machine-learned potential (MLP) is trained to approximate the PES of the Mg-H system, a well-known hydrogen storage material, to speed up the simulations and bring down the cost of PIMD. The data acquisition is accelerated by training the potential 'on-the-fly' during classical and path-integral molecular dynamics runs, where ab-initio calculations are replaced by the MLP whenever the estimated errors are low enough. In this contribution, I present the workflow for training and validation of this MLP, a process that is transferable to any metal-hydrogen system, and evaluate the speed-up that allows to perform computationally expensive PIMD calculations. First insights obtained from these calculations enhance our understanding of the Mg-H system on the atomic scale.

Monte Carlo Simulations of Water Pollutant Adsorption at Parts-per-Billion Concentration: A Study on 1,4-Dioxane

Samiha Sharlin, Rodrigo Lozano, Tyler R. Josephson

1,4-dioxane is an emerging water pollutant with high production volumes and a probable human carcinogen. The inadequacy of conventional treatment processes demonstrates a need for an effective remediation strategy. Crystalline nanoporous materials are cost-effective adsorbents due to their high capacity and selective separation in mixtures. This study explores the potential of all-silica zeolites for separation of 1,4-dioxane from water. These zeolites are highly hydrophobic and can preferentially adsorb nonpolar molecules from mixtures. We investigated six zeolite frameworks (BEA, EUO, FER, IFR, MFI, MOR) using Monte Carlo simulations in the Gibbs ensemble. The simulations indicate high selectivity by FER and EUO, especially at low pressures, which we attribute to pore sizes and shapes with more affinity to 1,4-dioxane. We also demonstrate a Monte Carlo simulation workflow using gauge cells to model the adsorption of an aqueous solution of 1,4-dioxane at 0.35 ppb concentration. We quantify 1,4-dioxane and water coadsorption, and observe selectivities ranging from 1.1×10^5 in MOR to 8.7×10^6 in FER. We also demonstrate that 1,4-dioxane is in the infinite dilution regime in both the aqueous and adsorbed phases at this concentration. This simulation technique can be extended to model other emerging water contaminants such as per- and polyfluoroalkyl substances (PFAS), chlorofluorocarbons, and others, which are also found in extremely low concentrations.

Molecular Models Indicate Stronger Binding of HIV-1 to Co-Receptor CXCR4 Compared to CCR5 at a Viral Population Level

Bailey Zinger, Jonathan Faris, Kayla G. Sprenger

In order for Human Immunodeficiency Virus Type 1 (HIV-1) to enter host cells, its "Env" surface glycoprotein engages a primary cellular receptor, CD4, and then one of two possible co-receptors, CCR5 or CXCR4. The amino acid sequence of HIV-1 Env is either CCR5-tropic, CXCR4-tropic, or dual-tropic. Transmitted/founder strains of HIV-1, which seed new human infections, are usually CCR5-tropic. However, as HIV-1 infection progresses, viral variants can arise that have CXCR4- or dual-tropism. We gathered nearly all sequences of Env in the Los Alamos National Laboratory HIV Sequence Database. We utilized an experimentally validated computational scoring function to calculate the free energy of interaction between each of these Envs and either CCR5 or CXCR4. We found that regardless of tropism, Env typically binds CXCR4 more strongly than CCR5. Conversely, CXCR4-tropic Envs bound to both co- receptors more strongly than CCR5-tropic Envs.

We further employed physics-based all-atom simulations of protein-protein interaction to calculate the free energy between each of these Envs and either CCR5 or CXCR4, in which the co-receptor was embedded in a lipid bilayer. We engineered CXCR4-tropic mutations in CCR5- tropic Env. We found that such mutations caused disruption of the stability of Env when bound to CXCR4 but not to CCR5.

Based on these results, we predicted that most CXCR4-tropic Envs bind CCR5 too strongly and most CCR5- tropic Envs bind CXCR4 too weakly for efficient infection. We predicted that an optimal degree of binding affinity and flexibility between Env and its co-receptor is required for efficient infection. The optimal affinity of Env for each co-receptor is different, explaining why not all Envs bind both CCR5 and CXCR4.

Ultimately, our results shed new insights into the sequence population-level thermodynamics and molecular-level mechanisms of HIV-1 infection, supporting further research into novel or improved therapeutics to treat people infected with HIV-1.

Self-Diffusion of Volatile Organic Compounds in Zr-MOFs: A Joint Computational and Experimental Study

Nickolas Gantzler, R Eric Sikma, Hannah Gruenwald, Michael E. Chandross, Keith Fritzsching, Dorina F. Sava Gallis, N. Scott Bobbitt

The association of specific volatile organic compounds (VOCs) in exhaled human breath (EHB), often called a chemical "fingerprint", to a particular disease is a growing field of study. The ability of trained scent dogs to identify COVID-positive respiratory secretion samples lends itself to the possibility of developing a non-invasive, breath-based diagnostic tool to provide insight into a person's state of health. However, detecting the target constituents --- often existing at dilute concentrations in a complex mixture of water, CO2, and other VOC interferents --- poses a significant challenge.

Metal-organic frameworks (MOFs) have demonstrated tremendous aptitude for applications in gas sensing and separation technologies. MOFs are porous, crystalline solids composed of metal nodes coordinated by organic linkers which form an extended network. Zirconium (Zr) MOFs exhibit exceptional thermal and hydrolytic stability which makes them attractive candidates for studying VOC detection in EHB. If MOFs are to be used in a beath-based diagnostic tool, understanding the interactions between the VOCs and the MOF is vital.

In this work, we use molecular dynamics simulations and experimental NMR to study the self-diffusion properties of several polar and non-polar VOCs in a topologically diverse set of Zr-MOFs. We compute self-diffusion coefficients at multiple loadings over a range of temperatures and compare against experimental values. We will discuss trends between self-diffusion and heat of adsorption, molecular properties (e.g. chain length, polar or non-polar, and isomers), and framework topology. Finally, we consider the challenges of comparing simulation to NMR experiments.

SNL is managed and operated by NTESS under DOE NNSA contract DE-NA0003525.

Monte Carlo Simulations with Feasst

Harold Hatch, Daniel Siderius, Vincent K. Shen

FEASST is open-source Monte Carlo simulation software maintained by permanent staff scientists at NIST (https://feasst.nist.gov). Methods and ensembles include Metropolis, Wang-Landau, Transition-Matrix and Mayer-sampling Monte Carlo in microcanonical, canonical, isothermal-isobaric, grand-canonical, semi-grand, Gibbs and expanded ensembles. Interaction potentials include square well, Kern-Frenkel, Lennard-Jones, Mie, Yukawa, Ewald and table potentials with cell and neighbor lists. Bonded potentials include rigid, square well, harmonic, and FENE bonds and angles with TraPPE and Ryckaert-Bellemans dihedrals. Confinement potentials include slab, cylindrical, spherical, table and 2D Ewald corrections. Trials include translation, rotation, crankshaft, pivot, rigid cluster, alchemical transformation, (dual-cut) configurational bias, aggregation volume bias, reptation and Jacobian-Gaussian branching. Written in C++ with interfaces to Python, server client ports, MPI and text file, FEASST supports checkpointing, restarting, postprocessing and OMP parallelization with flat-histogram and prefetching.

Application of Free Energy Landscape Interpolation and Extrapolation for Large-Scale Computational Screening of MOFs for Use in Water Adsorption

Bartosz Mazur, Lucyna Firlej, Bogdan Kuchta

Sorbent-assisted atmospheric water harvesting (AWH) has great potential for decentralised sourcing of drinking water in many arid regions of the world. Metal–organic frameworks (MOFs) offer great potential for this application because of their structural diversity and the possibility of their design and functionalisation. The total number of synthesised MOF structures is difficult to estimate – the CoRE MOF 2019 database contains ~14,000 MOF structures, while the CSD MOF subset contains a striking number of ~100,000 structures. Considering the number of existing and theoretically possible structures, to significantly reduce the cost in time and resources for the best materials selection the use of computational techniques validated by experiment is crucial.

Despite being the current state-of-the-art method, grand canonical Monte Carlo (GCMC) simulations for water adsorption have been reported as inefficient and prone to errors. The source of these problems lies in the mechanism of water adsorption in MOFs, which is overwhelmingly dominated by adsorption through the formation of water clusters that fill entire pores. In the course of the classical GCMC simulation, water clusters are difficult to form and break because of the large energy barriers between states. For this reason, we advocate the use of the grand canonical transition matrix Monte Carlo (GC-TMMC) method, which allows to calculate the free energy of the system, from which the adsorption isotherm can be determined. As the computational cost of the transition matrix method is high, we propose using the *NVT + ghost swap* method along with a new interpolation scheme. Our approach makes it possible to efficiently determine the full adsorption isotherm and distinguish between stable and metastable states, while dramatically reducing the time and number of required simulations. The use of the TMMC method allows for temperature extrapolation of the system's free energy profile. Thus, we can predict the adsorption properties of the system over a wide range of temperatures and pressures, on the basis of only a few simulations. This not only offers the potential to improve high-throughput screening studies in sorbent-assisted AWH using MOFs but also enhances the ability to optimise other processes requiring temperature/pressure swing adsorption.

Comparison of Diffusion and Flow Models on Biophysical Data

Richard John, Lukas Herron, Pratyush Tiwary

In biophysics and chemistry, generative models can augment small, difficult-to-obtain experimental datasets and output new, realistic data. Mathematically, their goal is to draw samples from the same probability distribution as some existing samples without knowing the underlying distribution's form. Diffusion and flow models are two popular categories of such models. Given unlimited resources, both diffusion and flow models could generate samples of perfect quality that strictly obey the unknown probability distribution. However, it would be helpful to know which model performs best in realistic circumstances. Here we perform numerical experiments to measure the sample quality of three generative algorithms with emphasis on molecular sciences: denoising diffusion probabilistic models (DDPM), normalizing flow models (NF), and continuous normalizing flow models (CNF). By selecting open-source implementations of each algorithm and using two realistic sets of data for testing, we provide a more specific and concrete comparison than existing stochastic framework-level approaches. Our datasets contain multimodal Gaussian distributions in high dimensions and molecular dynamics data. This selection of experiments is designed to mimic molecular data, establishing the pros and cons of different generative models. This opens the possibility of combining learned representations of molecular data in conjunction with other algorithms to accomplish goals such as predicting conformations of proteins or nucleic acid molecules such as RNA.

Calculation of Thermodynamic Properties of Helium By Path Integral Monte Carlo Simulations Using *Ab Initio* Potentials

Philipp Marienhagen, Karsten Meier

Based on Feynman's path integral formulation of quantum mechanics, nuclei can be treated quantum mechanically in a classical Monte Carlo simulation by representing particles as finite ring polymers within the classical isomorphism. In this work, the methodology of Lustig [1], with which rigorous expressions for all thermodynamic properties can be derived in any ensemble, is extended to the path integral framework for the isobaric-isothermal (*NpT*) ensemble. With the derived expressions, thermodynamic properties such as the speed of sound or Joule-Thomson coefficient can be calculated fully incorporating quantum effects without uncontrolled approximations. The derived expressions are verified by simulations of supercritical helium at selected state points at temperatures below 100 K using highly accurate *ab initio* potentials for pairwise and non-additive three-body interactions [2-3]. We observe good agreement of our results with recent experimental data [4] as well as the reference equation of state, which confirms our approach and demonstrates the necessity to consider quantum effects as well as nonadditive three-body interactions when predicting thermodynamic properties by solely theoretical means. We additionally investigate different simulation strategies including refined sampling techniques, such as the staging algorithm, and advanced methods to calculate thermodynamic properties based on virial estimators. Additional classical simulations enable the evaluation of the contribution of non-additive three-body interactions and quantum effects to different thermodynamic properties.

1. R. Lustig, J. Chem. Phys., **109**, 8816-8828 (1998); Mol. Sim., **37**, 457-465 (2011).
2. P. Czachorowski, M. Przybytek, M. Lesiuk, M. Puchalski, and B. Jeziorski, Phy. Rev. A, **102**, 042810 (2020).
3. J. Lang, G. Garberoglio, M. Przybytek, M. Jeziorska, and B. Jeziorski, Phys. Chem. Chem. Phys., **25**, 23395-23416 (2023).
4. T. Dietl and K. Meier (unpublished).

Comparing Scaling Relationships in Direct Catalytic Methane Conversion to Methanol

Kevin Yao, Randall Snurr

The current strategy for transforming natural gas into valuable chemical derivatives, such as methanol, involves a complex multistep process that requires high temperatures and pressures. The development of an alternative, direct catalytic method under mild conditions would be a significant breakthrough that would facilitate utilization of natural gas that is currently stranded in remote locations. A promising class of materials for this reaction are metal-organic frameworks (MOFs) containing open metal sites. Previous work has shown that there is an inverse scaling relationship between two important steps of methane conversion: the formation of a terminal metal-oxo group on the catalyst and the activation of the C-H bond in methane. This inverse scaling means that there is a trade-off between these two steps in the reaction pathway, and it is difficult to find catalysts that have low barriers for both. In contrast, investigations into single-site catalysts have revealed a less pronounced scaling relationship, prompting further investigation into breaking this scaling relationship.

To address this challenge, we conducted density functional theory calculations to predict the metal-oxo formation and the C-H bond activation energies, incorporating and expanding upon approximately 40 MOFs and 200 single-site catalysts that were studied in the literature [1, 2]. Our calculations show that structural distortions that have previously broken the scaling relationship in Fe (II) single-site catalysts are also promising for MOF clusters and other transition metals. We then imposed steric strain on potential MOFs in an attempt to break the scaling relationship. Understanding how this scaling relationship can be broken will help design new materials for efficient methane-to-methanol conversion.

[1] Andrew S. Rosen, Justin M. Notestein, Randall Q. Snurr. Structure–Activity Relationships That Identify Metal–Organic Framework Catalysts for Methane Activation. *ACS Catalysis* **2019**, *9* (4), 3576-3587.

[2] Gani, T. Z. H.; Kulik, H. J. Understanding and Breaking Scaling Relations in Single-Site Catalysis: Methane to Methanol Conversion by $Fe^{IV}{=}O$. *ACS Catalysis* **2018**, *8*, 975– 986.

Computational Analysis of the Thermodynamics of Ring-Opening Polymerization: Improved Entropy Predictions Via Cluster Extrapolation

Xiaoyang Liu, Shivani Kozarekar, Alexander Shaw, Linda J. Broadbelt

The potential of chemical recyclability and/or biodegradability holds promise for addressing the current plastics dilemma.[1,2] The major challenge in chemical recycling is the inefficient depolymerization of certain polymers. Yet, a distinct category of polymers, created through cyclic monomer opening, offers a notable avenue for depolymerization. Consequently, enhancing ring-opening polymerization (ROP) processes is crucial for sustainability efforts.[3] The Gibbs free energy of polymerization, denoted as $\Delta G_p = \Delta H_p - T\Delta S_p$ (where T is temperature, and ΔH_p and ΔS_p represent the enthalpy and entropy of polymerization, respectively), is a fundamental concept. The ceiling temperature (T_c), defined as the temperature at which polymerization and depolymerization rates are balanced, such that $\Delta G_p = 0$ and $T_c = \Delta H_p/\Delta S_p$, is an essential concept in designing new monomers for the creation of circular polymers.[2,4]

While it is possible to experimentally determine ΔH_p and ΔS_p, the *in silico* prediction of these values for high-throughput analysis presents a compelling yet challenging prospect. This challenge has impeded our progress towards the goal of designing polymers that can be depolymerized to achieve desired T_c values alongside other beneficial properties. Tran et al.[5] have recently introduced a method for calculating ΔH_p in ring-opening reactions using size-dependent loop structures to mimic finite-size polymer models, eliminating the distortions caused by varying end groups. Nevertheless, this method falls short in estimating ΔS_p, as it requires an accurately simulated solvent environment, which gas-phase calculations fail to provide.

In this research, we introduced a comprehensive workflow towards a high level of accuracy for calculating both ΔH_p and ΔS_p. This method incorporates the variations in the local solvation environments encountered during monomer polymerization across various solvents, including cases in bulk phase, where the monomer itself acts as the solvent. Our process utilizes sophisticated extrapolative models for monomers and oligomers, complemented by thorough cluster exploration of their configuration spaces via *ab initio* molecular dynamics (AIMD) simulations paired with density functional theory (DFT) calculations. Our results show that using the cluster extrapolation method, with the monomer surrounded by itself, minimally affects polymerization enthalpies but enhances the accuracy of polymerization entropy simulations by about 30 cal/mol·K in the bulk phase. These computational insights aim to aid, guide, and streamline the extensive design process of circular polymers.

References

1. Borrelle, S. B. *et al. Science* **2020,** *369* (6510), 1515–1518.

2. Shi, C. *et al. Chem* **2021,** *7* (11), 2896–2912.

3. Tardy, A. *et al. Chem. Rev.* **2017,** *117* (3), 1319–1406.

4. Liu, X. *et al. Cell Rep. Phys. Sci.* **2024,** 101910.

5. Tran, H. *et al. J. Phys. Chem. Lett.* **2022**, *13* (21), 4778–4785.

Why Classical Force Fields Are Unable to Predict Correct Water Dynamics in Salt Solutions

Nupur Mehra, Aditi Seshadri, Shuwen Yue

Predicting accurate dynamic properties of aqueous electrolyte solutions using molecular simulation is essential to understanding molecular scale mechanisms in countless domains, from electrochemistry to biology. Adding different salt species to water causes significant changes in the arrangement of water molecules. Kosmotropic salts (e.g., NaCl, NaF, NaBr) facilitate the formation of a structured network of water, leading to decreased water diffusivity. On the other hand, chaotropic salts (e.g., KCl, KBr) disrupt the hydrogen bond network, enhancing water diffusivity. Classical molecular dynamics simulations using empirical force fields fail to accurately capture these diffusion trends, indicating the absence of underlying physics. In previous work, we have shown that machine-learned potentials (MLPs) fit to a first-principles derived potential energy surface can predict the correct diffusion trends. In this work, we seek to identify why. To determine the underlying physics absent in classical force fields and present in first-principles MLPs, we systematically analyze the impact of polarizability, bond flexibility, and charge transfer on the predictive accuracy of water dynamics in aqueous salt solutions. The insights from this work serve as a basis for refining predictive models and understanding the fundamental knobs that control electrolyte solution properties.

Decoding Protein Structure and Hydropathy

Jingjing Ji, Shikha Nangia

The hydrophobic effect is crucial for many biomolecular interactions, including protein folding, protein-protein association, protein-ligand binding, and micelle formation. Among these, protein-protein interactions are significant for most physiological and pathological processes, such as immune response, signal transduction, and enzyme catalysis. The protein-protein interaction interfaces involve hydrophobic patches, which are influenced by the chemistry of the underlying amino acid residues and depend on the surface patterns. The previously reported approaches either fail to capture the chemical and geometrical patterns of the heterogeneous surface or are computationally expensive. We have developed a new computational method and protocol for Assigning a Residue's Character on a Hydropathy (PARCH) Scale. It provides the collective response of water to the topological features of the proteins. We investigated an array of proteins and identified hydrophobic patches of proteins that lead to protein assembly. The results are in excellent agreement with those published in the literature. Our results show that amino acids exhibit disparate hydration properties depending on the local topography of the protein. In addition, this method enables us to compare the hydrophobicity of different proteins quantitatively relative to each other. After that, we evaluated the hydropathy of 277,877 amino acids in 1,000 structurally different proteins using PARCH. Interestingly, the data revealed the presence of earth-like zones spanning each protein's shell, mantle, and core. The outermost shell is enriched in positively charged lysine and arginine residues, while the core is dominated by leucine. The intermixing of the shell and mantle residues gives rise to the protein's rich structural heterogeneity and, in turn, its hydropathy. Using the core-shell-mantle construct, we show how positions and neighborhoods of residues modulate their surface properties, such as hydrophilic and hydrophobic pockets and bumps at the nanoscale.

Extending Mosdef Tools for the Reproducible Investigation of Thermoplastic Weld Strength in Aerospace Composites

Chris Jones, Marjan Albooyeh, Rainier Barrett, Eric Jankowski

Large structures made from thermoplastic composites can be engineered by welding together pieces that are mechanically joined on the nanoscale through chain entanglements across the weld interface. Understanding how polymer chemistry and processing effect weld strength is therefore important to reliably manufacturing aircraft components such as rotor blades, wings, and fuselages. In this work we develop computational infrastructure for studying such systems: the Flexible Library of Organic Workflows and Extensible Recipes for Molecular Dynamics (FlowerMD). This library extends the Molecular Simulation Design Framework and exemplifies the capabilities in complex workflow management that derive from using community-developed tools: Initializing molecules and parameterizing force fields of detailed models that are used to provide reference trajectories from which machine-learned coarse potentials are derived. After validation coarse-grained models are then used in FlowerMD to perform fusion welding simulations. We demonstrate validated models of poly(phenylene sulfide) (PPS) and poly(ether ketone ketone) (PEKK) as case studies demonstrating the general capabilities of FlowerMD.

Designing Selective Nanoporous Materials for VOC Capture Applied to Breath Diagnostics: Insights from Simulation and Experiments

Scott Bobbitt, R Eric Sikma, Michael E. Chandross, Jason Sammon, Dorina F. Sava Gallis

There are numerous volatile organic compounds (VOCs) present in exhaled human breath. Different diseases or metabolic states produce unique profiles of some of these VOCs, providing a chemical fingerprint that has potential to be used as a diagnostic tool or provide insight to the health of the body. However, detecting these specific molecules in a complex mixture of air, water, CO_2, and other non-marker VOCs is a significant challenge. Metal-organic frameworks (MOFs) are porous crystalline materials consisting of inorganic nodes connected via organic linkers. MOFs offer a variety of pore sizes, pore shapes, and chemical properties, implying that they can be tailored to specific adsorption applications. We use high-throughput molecular simulation to design or choose sorbents tailored to adsorb specific target VOCs with high selectivity over competing molecules (e.g. water, CO_2). We also use simulation to find MOFs suitable for difficult isomer separations. We have identified and synthesized several promising MOFs with high selectivity for VOCs of interest—including some known to be indicators for COVID-19 -- and validated their adsorption capacity and selectivity in realistic synthetic breath streams.

We will also discuss diffusion properties of selected VOCs in several MOFs of interest and experimental tests of the chemical and thermal stability of these MOFs in humid environments.

SNL is managed and operated by NTESS under DOE NNSA contract DE-NA0003525

Molecular Modelling of Hydrogen Behavior in Vapor-Liquid Equilibria and Confinement in Carbon-Based Materials.

Andrés Mejía, Harry Cardenas

With the imminent migration to greener energy carriers, the use of hydrogen seems like the most plausible option in the near future. Regardless of the method to obtain the hydrogen, the main issue that needs attention is how to store it in an efficient way. One of the options that has been highly studied over the years corresponds to the confinement of the hydrogen molecules into nanoporous materials.

Across the several studies available related to adsorption of hydrogen, different materials, and conditions have been employed to find the optimal combination that brings efficiency to the process. The most preferable route is adsorbing the hydrogen molecules at cryogenic conditions (77K) in order to maximize the sorption capability due to the low temperature. However, the small size of the hydrogen molecules, the high density induced by the confinement, and the low working temperature make the quantum effects significant. In addition to the accurate characterization of the fluid molecules, the quantum treatment of hydrogen molecules is needed to achieve the agreement between the experimental data available and the molecular models (both theoretical and molecular simulations). Among the alternatives available to calculate thermodynamic properties for small molecules with quantum characteristics such as hydrogen, the Feynman-Hibbs (FH) correction to the intermolecular potential is a reliable approximation to consider these quantum effects.

The present work employs the Feynman-Hibbs corrections to the intermolecular potential to perform Monte Carlo simulations. These are used to assess the effectiveness of the FH quantum corrections through the calculation of the vapor-liquid equilibria (VLE) and the adsorbed amount under confinement in carbon-based materials. Besides the corrections to the fluid-fluid interactions in the VLE and adsorption calculations, the FH corrections are applied to the fluid-solid interactions due to the solid walls of the material to evaluate their importance at the confined level. Moreover, different carbon-based materials are explored to see how the variations in their structural morphology can affect the resulting isotherms. Among the structures studied, we can mention carbon nanotubes, graphene slit-like pores, corrugated graphene slit-like pores, graphane slit-like pores, graphene nanoflakes amorphous materials, graphane nanoflakes amorphous materials, amorphous materials built from platonic hydrocarbons (tetrahedrane, cubane and dodecahedrane), and fullerene-based materials.

Multiscale Responsive Kinetic Model (MsRKM): Capturing the Impact of Electrochemical Gradients on Channels and Transporters

Hannah Weckel-Dahman, Jessica M. J. Swanson

The active transport of ions across the cell membrane creates an electrochemical gradient that plays a central role in various biological processes. The resulting transmembrane potential combines a chemical force (driven by a concentration gradient) and an electrical force (driven by a charge imbalance) to regulate transmembrane transport. Previous work with Multiscale Kinetic Models (MKM) revealed flux through heterogeneous pathways shift in response to different chemical gradients, lending credence to the idea that kinetic selection plays a role in membrane transport. In this work, we present Multiscale Responsive Kinetic Modeling (MsRKM), a framework that Integrates data from multiscale simulations and experiments like MKMs while also allowing the kinetic network to respond to different electrochemical conditions. We demonstrate MsRKM on coupled Cl^{-}/H^{+}ion exchange in the chloride channel antiporter, CLC-ec1, and on the SHAKER-K^{+} channel, known for rapid ion transport. We show how the physically unique influences of the electrical and chemical gradients produce different mechanistic pathways and environmental responses. More generally, we demonstrate how MsRKM allows us to explore how proteins respond to different reaction conditions and predictively model mutant properties when the substrate binding site stability change is known. Additionally, the discovered asymmetry in ClC-ec1's response to electrochemical gradients suggests a potential role in *E.coli* extreme acid response.

Multiscale Modeling and Simulation of Defect-Engineered Zirconium Metal Organic Frameworks (MOFs) for CO_2 Capture

Muhammad Hassan, Yu Chen, Yongchul Chung

The presence of defects in the crystalline structure of metal organic frameworks (MOFs) is known to influence the properties of the material. Defect engineering of MOFs offers a promising strategy for fine-tuning their properties, such as the shape of adsorption isotherms, leading to a potential improvement in performance for specific applications, such as CO_2 capture. In this study, we develop and optimize several defect structures of Zr-MOFs (UiO-66, UiO-67 and MOF-801) with varying defect concentrations ranging from 0-15%. We carry out grand canonical Monte Carlo (GCMC) simulations to generate CO_2 and N_2 isotherms. We then perform process modeling and optimization based on the isotherm data to investigate the correlation between defect concentration and material's performance at the process-level. The comparison of pareto front curves between purity and recovery and the distribution of optimal decision variables is then used to correlate the degree of defects and the process performance.

Machine Learning Methods to Tackle the Problem of RNA-Drug Binding

Venkata Sai Sreyas Adury, Pratyush Tiwary

Contemporary medicine predominantly revolves around drug-protein interactions as the principal therapeutic approach, limiting the scope of medicine to easily druggable protein targets, neglecting over 85% of proteins in our body. Messenger RNAs (mRNAs) serve as templates for protein synthesis, and many non-coding RNAs are also implicated in diseases, motivating us to target RNA instead.

Unlike for proteins, far fewer crystal structures are available for RNA. Moreover, RNA exhibit a diverse ensemble of structures in solution, undermining the completeness of crystallographic insights even when structures are available. Thus, we lack both a significant number of RNA-binding drug molecules and a substantial repository of RNA structures. Machine Learning methods have recently shown great promise with such problems, but their extension to RNA has been poor.

In this presentation, we'll explore machine-learning methods to address some of these challenges.

Controlling Topological Defects in Curved Spherical Crystals of Hard Superballs

Gabrielle Jones, Philipp Schönhöfer, Sharon Glotzer

Much like a soccer ball cannot be manufactured solely with hexagonal patches, spherical particles cannot decorate sphere surfaces in a hexagonal lattice without defects. In physical systems such as Pickering emulsions, which are stabilized by the addition of colloids to a droplet interface, the icosahedral arrangement of dislocations and disclinations on the surface dictates their mechanical and optical properties. Here we use hard particle Monte Carlo simulations to determine the effect of shape anisotropy on underlying lattice preferences and defect structures on a closed sphere surface. By analyzing a shape family of cubic superballs on spherical surfaces we demonstrate that cuboidal particles form a simple square assembly and overcome the topological incompatibility by distributing eight three-fold defect regions evenly about the surface. Additionally, by varying particle superball shape smoothly from cubes to spheres we unveil how the global defect motif transitions from cubic to icosahedral symmetry. Congruent studies of tetrahedral superballs reveal additional varieties of simple square, honeycomb, and hexagonal motifs. We additionally describe the differences in defect morphology for each packing motif. This work has promising implications for the stability of Pickering emulsions as well as access to different vesicle buckling modes such as those seen in icosahedral virus capsids.

Acknowledgment: 'This material is based upon work supported by the U.S. Department of Energy, Office of Science, Office of Advanced Scientific Computing Research, Department of Energy Computational Science Graduate Fellowship under Award Number DE-SC0022158.'

Modeling Temperature-Dependent Transport Properties in Dissipative Particle Dynamics: A Top-Down Coarse-Graining Toward Realistic Dynamics at the Mesoscale

Nunzia Lauriello, Gianluca Boccardo, Daniele Marchisio, Martin Lisal, Antonio Buffo

Soft materials play pivotal roles in industries such as food, personal care, and pharmaceuticals. The rational design and performance optimization of such materials and equipment involved in their processing require the ability to predict their morphology and their properties on process-relevant length and time scales in a reasonable time. However, this task is hindered by time-consuming and costly experimental campaigns needed to estimate material properties, which are influenced by numerous variables such as temperature. Predicting the concurrent changes in morphology and properties induced by temperature variations poses a significant theoretical challenge. In this context, computational tools offer an intriguing alternative: they facilitate the mimicking of various temperature conditions rapidly and cost-efficiently, providing insights into the complex interplay between temperature and material properties. Among the available computational tools, Dissipative Particle Dynamics (DPD) is one of the most applied to explore the behaviour of soft materials and estimate their properties. It is a mesoscopic technique that operates at time and length scales relevant to capture the evolution of the morphologies that dictate the final soft materials properties. The reason for this is that DPD is a coarse-grain (CG) model, where groups of atoms and molecules are treated as single entities called beads, interacting via a mesoscopic force field, made of conservative soft-repulsive and non-conservative interactions. However, there are still some gaps in the method which narrow its scope of application. For example, since there are long standing ambiguities regarding the introduction of the system temperature variation into the model, the existing DPD research typically overlooks its effects by operating at fixed system temperature. Therefore, this research aims to incorporate the effect of system temperature variation within the DPD model to capture realistic temperature-dependent properties. This work focuses on transport properties, which are often overlooked despite their significant influence on the flow dynamics of non-isothermal mesoscopic systems. Moreover, before simulating the response of the system induced by a temperature change, it is important to estimate transport properties at a certain temperature. In this work, we investigated the use of an "extended" DPD model, which was proved to be suitable for the description of realistic transport properties unlike the standard one, and thus to predict temperature dependent transport properties at the mesoscale following a top-down CGing. We made use of an in-house developed automated method (https://github.com/mulmopro/AMTP_DPD) to compute transport properties from equilibrium simulations, which allowed us to obtain estimates of these quantities with controlled accuracy, adopting a systematic and reliable computational protocol. In particular, we identified a temperature-dependent parametrization methodology for both conservative and non-conservative interactions capable of ensuring the correctness of not only equilibrium properties but also dynamic properties. In fact, we successfully validated this methodology against experimental data of water transport properties among the whole liquid temperature range. Furthermore, we presented the results in the context of a critical comparison with theoretical relationship focusing on the temperature effect. We found that the findings are promising in the field of mesoscale modeling of non-isothermal systems.

Role of Lipids in Methane Mass Transfer to Particulate Methane Monooxygenase (pMMO)

Pooja Rao, Gabriel Costa Alverni da Hora, Jessica M. J. Swanson

Methane is a potent global warming gas with a short-term warming potential 86 times that of CO2, instigating an urgent need for effective methane removal strategies. Recently, the potential use of microbial enzymes in methane mitigation has been explored as an environmentally and economically friendly alternative to chemical counterparts. Particulate methane monooxygenase (pMMO) is one such microbial membrane-bound enzyme present in methanotrophs, capable of oxidizing methane into methanol. The main challenges that prevent immediate utilization of pMMO for methane removal are low methane consumption rates at near ambient air levels and the inability to retain pMMO activity outside the native environment. It is hypothesized that native lipids in the pMMO system aid in methane mass transfer and enable oxidation activity by influencing the structural and dynamic properties of pMMO. However, the exact identity and mechanisms by which the native membrane environment influences the activity of pMMO remain unknown. To bridge this knowledge gap, we aim to understand methane uptake and permeation rates in model membranes of varying lipid composition and identify the native lipids in the cryogenic electron microscopy (cryo-EM) structure of pMMO. We have utilized molecular dynamics (MD) simulations to determine the partition coefficients of methane between model membranes and water via probability and free energy distributions. To accurately determine the rates of methane entrance and escape in model membranes, we have employed a path sampling technique, namely replica exchange transition interface sampling (RETIS). To characterize the lipids in the vicinity of the cryo-EM structure of pMMO, we applied LipIDens, a pipeline integrating simulations to interpret lipid densities. Initial data from our simulations indicate membrane composition and position-dependent partition coefficients, and permeation rates for methane. Ultimately, these studies will guide the development of pMMO-based technologies that can efficiently consume methane from near ambient air levels.

Modeling Nucleation of Molecular Crystals from Solution

Jacob McKibbin, Erik Santiso

Nucleation from solution is ubiquitous in nature, and frequently appears in applications in science and engineering, including drug and food manufacturing, separations and the manufacture of electronic components. Despite this, direct simulation of nucleation from solution is challenging to study. Nucleation is a rare event - it is unlikely to be observed in the timescale of a typical simulation. Such systems require specialized 'rare event' techniques to observe the transition. In this work, we use one such technique, the string method in collective variables, along with the Santiso-Trout order parameters for molecular crystals to calculate the free energy barrier to nucleation from solution. We studied the model drug sulfadiazine at five different levels of supersaturation in acetonitrile. We found that the free energy barrier to nucleation decreases monotonically with increasing supersaturation, as does the size of the critical nucleus. We drew from classical nucleation theory, and found a heuristic rule in terms of saturation, and surface energy which was in good agreement with our simulations. Finally, we developed an open source module implementing the aforementioned order parameters into PLUMED which, along with our in-house suite of code, can be used for free energy calculations via the string method in collective variables on any platform supporting PLUMED.

Adsorption in Flexible Soft Porous Coordination Polymers

James Carpenter, Yamil Colón

Soft porous coordination polymers (SPCPs) are materials with exceptional potential because of their ability to incorporate the properties of nominally rigid porous materials like metal–organic frameworks (MOFs) and those of soft matter, such as polymers of intrinsic microporosity (PIMs). This combination offers the gas adsorption properties of MOFs together with the mechanical stability and processability of PIMs, opening up a space of flexible, responsive adsorbing materials. We examine the adsorption properties of these materials using a combination of Grand Canonical Monte Carlo simulations in RASPA and molecular dynamics in LAMMPS. Further, we examine the effects of material building blocks and structure on adsorption, and on resulting pore size distribution and radial distribution functions.

Insights from Molecular Simulation on Decrystallization of Polyesters

Daria Lazarenko, Brandon C. Knott

Utilizing waste polyesters as raw materials for recycled goods is an important sustainable strategy. The semi-crystalline nature of these materials complicates their depolymerization, making it crucial to examine the energy required to decrystallize polyesters. This study proposes a methodology employing molecular dynamics simulations to determine the energy needed to detach a single polymeric chain from the crystalline surface of five key polyesters (PET, PTT, PBT, PEN, PEF) in water. The results indicate a substantial variation in energy requirements, with PEN needing about 15 kcal/mol and PEF only 8 kcal/mol per monomeric unit. This research sheds light on the molecular interactions within the structure of semi-crystalline synthetic polyesters and offers valuable insights for experimentalists aiming to enhance plastic recycling efficiency through improved enzyme selections, as well as catalysis, recycling conditions, and material design.

Unusual Adsorption and Structural Transformations Mechanisms in Ordered Micropores

Bogdan Kuchta, Bartosz Mazur, Lucyna Firlej, Malgorzata Stankiewicz

The phase behavior of confined fluids adsorbed in nanopores differs significantly from their bulk counterparts and depends on the chemical and structural properties of the confining structures. In general, phase transitions in nanoconfined fluids are reflected in stepwise adsorption isotherms with a pronounced hysteresis. Here, we show experimental evidence and an in-silico interpretation of the reversible stepwise adsorption isotherm which is observed when methane is adsorbed in the rigid, crystalline metal– organic framework IRMOF-1 (MOF-5) [1]. In a very narrow range of pressures, the adsorbed fluid undergoes a structural and highly cooperative reconstruction and transition between low-density and high-density nanophases, as a result of the competition between the fluid–framework and fluid–fluid interactions. This mechanism evolves with temperature: below 110 K, a reversible stepwise isotherm is observed, which is a result of the bimodal distribution of the coexisting nanophases. This temperature may be considered as a critical temperature of methane confined to nanopores of IRMOF-1. Above 110 K, as the entropy contribution increases, the isotherm shape transforms to a common continuous S-shaped form [2] that is characteristic to a gradual densification of the adsorbed phase as the pressure increases. The other IRMOF structures have been also modeled and the results compared to the IRMOF-1 case. The density distribution shows symmetric structures which structural properties evolve in a function of uptake and temperature [3].

References:

1. B. Mazur, F. Formalik, K. Roztocki, V. Bon, S. Kaskel, Alexander V. Neimark, Lucyna Firlej, and Bogdan Kuchta, Quasicontinuous Cooperative Adsorption Mechanism in Crystalline Nanoporous Materials, J. Phys. Chem. Lett. 2022, 13, 6961–6965
2. D. Fairen-Jimenez, N. A. Seaton, T. Düren, Unusual Adsorption Behavior on Metal-Organic Frameworks. Langmuir 2010, 26 (18), 14694–14699.
3. Kuchta, B.; Dundar, E.; Formalik, F.; Llewellyn, P. L.; Firlej, L. Adsorption-Induced Structural Phase Transformation in Nanopores. Angewandte Chemie - International Edition 2017, 56 (51), 16243–16246.

Acknowledgements:

Supported by the National Science Centre (NCN), Poland (grant: 2022/45/B/ST8/02028)

Computing Noble Gas Solubility in Alkali Halide Molten Salts

Cole Strickling, Edward Maginn, Yong Zhang

In this study, we addressed the solubility of noble gases in molten alkali halide salts, crucial for the management of the fuel salt cycle in molten salt reactors (MSRs). Understanding the interactions of fission products, especially xenon and krypton, with spent fuel salts is vital for MSR efficiency and safety. We combined molecular dynamics simulations, conducted with LAMMPS and CP2K, and Monte Carlo simulations, using Cassandra, to predict Henry's law constants for the interactions between molten salts (NaCl and KCl) and noble gases (Ar and Xe) through the application of the Widom insertion method.

Our application of the Widom insertion method to systems of molten NaCl and KCl with Ar and Xe successfully yielded Henry's law constants and enthalpy of solvation values. These findings align with indirect experimental data, underscoring the potential of the Rigid Ion Model (RIM) for accurately predicting gas solubility in molten salts. A notable advancement of our research is the enhancement of RIM parameter accuracy through *ab initio* simulations, offering an improvement over traditional methods reliant on the Lennard-Jones potential.

Our study navigated through challenges, particularly the limited direct experimental comparisons, to advance our grasp of noble gas solubility in molten salts. This study represents a crucial development in understanding the interactions of noble gases within alkali halide molten salts, contributing fundamentally to the solute-solvent interaction science important for MSR operation and safety.

Microkinetic Modeling Meets Macroscopic Corrections

Jessica M. J. Swanson, Ryan Carlsen, Hannah Weckel-Dahman

Kinetic selection has a significant impact in biology, from ribosomal proofreading to substrate transport across membranes. Understanding the origins of kinetic selection requires the development of effective tools for modeling both steady-state and time-dependent behavior of reaction networks. However, most kinetics software is designed from the chemical reactor perspective, creating a significant gap in its application at the biomolecular level. Our software seeks to bridge the gap by creating a clear pipeline focused on the methods common to the field of biophysical chemistry. First, we obtain structural information to define intermediate states and establish a reaction network from experiments and molecular dynamics. Second, we quantify transition rates (defining the network edges) from multiscale simulations such as reactive, polarizable or classical molecular dynamics simulations combined with enhanced sampling or Brownian dynamics for diffusive substrate uptake. Finally, we refine the calculated rate coefficients to be consistent with electrochemical experiments. This process allows us to clearly link microscopic parameters with macroscopic observables in a single reaction network, improving our understanding of the mechanistic underpinnings of kinetic selection. We apply this method to the development of accurate time resolved-models for channel proteins.

Developing an Open-Source Tool for Generating Rich and Consistent Sigma Profiles

Fathya Salih, Dinis Abranches, Yamil Colón

The field of computational material discovery using machine learning has seen a lot of development in terms of novel neural network architectures, advanced training protocols, and novel molecular descriptors. However, the latest of these has an unsolved issue. All the commonly used molecular descriptors (SMILES strings, graphs, positional coordinates, etc.) share the problem of being dependent on molecule size. However, for most deep learning models, the size of the input has to be fixed to the size of the largest molecule in the dataset. Meaning, the network will end up with many unnecessary parameters and will require a larger training dataset. Additionally, the neural network that may have taken months to train will have an upper limit for the size of molecule it can be used with in the future.

As an alternative, sigma profiles have been proposed as a universal molecular descriptor. A sigma profile is an un-normalized histogram of the surface screening charge distribution of a molecule when embedded in a continuum solvent with a very high dielectric constant. Therefore, sigma profiles contain valuable information about the possible inter- and intra-molecular interactions the molecule can experience. And they have been successfully used as inputs to convolutional neural networks and gaussian processes to predict bulk material properties (e.g. boiling point, density, aqueous solubility, etc.) with minimal datasets of less than 1500 molecules. Furthermore, because they are histograms, their size is independent of molecule size - the sigma profiles for methane and C30 will both have the same number of bins. The main reason sigma profiles have not been used as molecular descriptors in more machine learning applications so far is the lack of an open-source tool for generating them, and the publication limits placed by commercially available tools.

This work summarizes the development of an open-source python tool for generating sigma profiles to allow for their use in large-scale materials discovery. The work, also, addresses the effect of conformers on sigma profiles to ensure the generated profiles are consistent and rich with information.

Physics-Informed Machine Learning for Biological Systems

Anjali Verma, Dedi Wang, Lukas Herron, Pratyush Tiwary

Machine learning methods provide a novel approach to analyzing many biological systems, including protein and nucleic acid folding processes. In this work, we employ a dynamics-constrained representation learning framework to study the conformational dynamics of small synthetic peptides alanine dipeptide and AIB9. Furthermore, by building a Markov model in the low-dimensional latent space and correcting for the fractional occupancies of states, we are able to learn meaningful dynamical information from relatively short trajectories. Extending this work to nucleic acids, specifically RNAs, is nontrivial as a simple representation learning method is not capable of capturing the glassy landscape of RNA folding. To tackle this problem, we look to integrate diffusion models within our dynamics-constrained framework.

A Modular Ecosystem for Atomistic Machine Learning

Guillaume Fraux

Machine learning is imposing itself more and more in the field of atomistic simulations, being used everywhere from large dataset analysis, direct prediction of various physical quantity, replacement for *ab initio* calculations and finds new applications every day. This explosion of applications comes with an explosion of software implementations for all the associated methods, typically isolated from each other and unable to be used together. This makes it a lot harder to combine existing methods and ideas for new science, or even just to compare existing methods for a given task.

We are working on **metatensor**, a free and open source software library to enable communication between different ecosystems, and make it easier to incorporate multiple building blocks in a single pipeline. This is made possible by augmenting the data created by one package with all the required metadata so another package can interpret and use the data. This metadata also enables us to exploit the inherently sparse structure of atomistic machine learning data, saving on memory usage and compute time. On top of this fundamental data format, we provide tools to construct new machine learning models, and provide a unified interface to train many different machine learning architectures, including architectures written without the knowledge of metatensor.

Finally, we are defining a minimal interface between machine learning models and simulations engines (such as LAMMPS, ASE, GROMACS, CP2K, PLUMED, ...) giving model authors the ability to use any engine by following this simple interface; and giving model users the ability to use the same models in many end to end workflows. Overall, we hope that this library will help create a modular ecosystem for atomistic machine learning, facilitate the testing of new idea by cutting redundant work; and enable non-machine learning experts to use it where relevant in their research.

Prediction of the Interfacial Tension and Micellization of Sugar-Based Surfactants from Molecular Simulations Using a Coarse-Grained Force Field

Ariif K. Bahrin, Omar Matar, Erich Muller

Sugar-based surfactants (SBS) are "green" amphiphiles with potential applications in various industrial settings, including pharmaceutical products, and personal and household care. The large industrial-scale production of these surfactants has become possible due to the availability of feedstocks synthesized from biomass through biorefineries. As opposed to the more common petrochemical surfactants (e.g., linear alkylbenzene sulfonates, fatty alcohol ethoxylates), SBSs are carbon-neutral, non-toxic, potentially cost-effective, and can be suitably tailored by modifying their morphology.

A typical SBS is composed of a polar head derived from glucose or maltose, a linker and a hydrocarbon (alkane) tail. Although molecular dynamics has been used and has achieved remarkable success in describing thermodynamic and molecular properties of systems with SBS, the traditional all-atom molecular simulation force fields considerably constrain the system size and the time scale of simulations. We employ coarse-grained intermolecular models parametrized via a top-down technique in which an accurate equation of state is used to link experimentally observed macroscopic properties of fluids with the force-field parameters. The state-of-the-art version of the statistical associating fluid theory (SAFT) for potentials of variable range as reformulated in terms of the Mie potential is employed here. With it, we propose a suite of coarse-grained (CG) model for alkylpolyglucoside, nonionic surfactants commercially known as APGs.AA molecular modelling approach is presented with a focus on predicting the surface tension and exploring the adsorption of the APGs on the water/air, water/n-decane and water/CO_2 interfaces. Canonical CG Molecular Dynamics (MD) simulations are performed, including both phases and the interfacial region with different levels of surface coverage up to and exceeding the critical micelle concentration.

Adsorption data obtained from the MD simulations, relating the interfacial tension, γ, (or surface pressure, Π) with surfactant surface coverage, Γ, are used to build a two-dimensional (2D) simulation-based equation of state. The equation of state is used to derive a model for the surfactant excess chemical potentials at the interface. The adsorption isotherm of APG_{12} (the γ – bulk concentration, c, curve) is determined from the derivation of a simple thermodynamic relationship that incorporates the calculation of the free energy of transfer transferring a surfactant molecule from the aqueous phase, through the interface and into the supernatant phase. These results are compared favorably to available experimental data, thus validating the models.

More interestingly, the CG nature of the model allows the exploration of large time and size scales that allow the observation of the detachment of micelles from super-saturated interfaces. Histograms of the micellar sizes and distributions are shown and compared across the different interfaces. The results reveal that the SAFT-based CG models are both robust and representative, having a higher accuracy than benchmark atomistically-detailed models but allowing the exploration of surfactant behaviour across a wide concentration range.

Inverse Design of Sbps Based on Sequence-Binding Relationships

Aditya Sonpal

Interaction between biomolecules and solid surfaces plays a vital role in designing new biomaterials and applications that mimic nature (e.g., biomineralization). Solid-binding peptides (SBPs) have gained significant attention due to their tunable sequences, which result in programmable binding energies. However, design of new SBPs is a non-trivial task driven by the relationship between the SBP sequence and the binding energy. This relationship is not deterministic from the amino acid (AA) sequence alone. This can be illustrated by Car9 (DSARGFKKPGKR). According to previous studies, with five positively charged AAs, Car9 binds strongly to a quartz surface (with $-OH^-$ ions). However, not all of them participate in binding. The same has been found for a few selected mutants or sequence isomers of Car9. Further, Car9 and its sequence isomers have conflicting binding behaviors to the quartz surface at different pH values. Therefore, we use Car9 and its ~20 million possible sequence isomers to study sequence-binding relationships. We narrow down this search space using unsupervised and semi-supervised clustering algorithms to end up with 1000 clusters with centers corresponding to 1000 SBPs, which form a reasonable representation of the Car9 isomer space. Here, we use the Levenshtein (or edit) and hamming distances as similarity (or dissimilarity) metrics between our SBP sequences and embed them into the clustering algorithm. We rank the 1000 resulting sequences based on their similarity with Car9. We perform high throughput computational screening of the 1000 resulting SBPs using a new method of rapidly estimating binding energies developed by our group. We then determine if sequences 'similar' to Car9 have similar binding energies to the quartz surface. This would be a starting point for creating design rules for SBPs based on physics-agnostic similarity metrics. Further, we rapidly estimate the binding energies of the SBPs at a higher pH and compare the behavior of the SBPs at two different pH values. The goal is to find which Car9 isomers will bind to the surface better at a higher pH and determine if we can attribute this behavior to the ordering of the AA sequence. Understanding these relationships between the AA sequence and binding behavior is a step toward the inverse design of SBPs.

First Principles Study of Mxene-Based Heterostructure TiS_2 /$Ti_3C_2O_2$ As Anode Materialfor Rechargeable Lithium-Ion Batteries

Aseres Arega

Energy storage technologies are critical for a broad range of applications, and lithium-ion batteries have emerged as a dominant and scalable electrochemical storage technology due to their high specific energy, coulombic efficiency, and number of charge/discharge cycles. However, to meet the increasing demand for energy storage, it is necessary to reduce the cost and increase the efficiency of these technologies. MXenebased materials are a novel class of 2D materials that have been developed into effective energy storage materials through years of research. In this study, we performed first-principles calculations to investigate the MXene-based heterostructure, T S_2 /$Ti_3C_2O_2$, as an anode material for rechargeable lithium-ion batteries. We employed density functional theory (DFT) with the PBE exchange-correlation functional and Grimme's DFT-D2 scheme to account for van der Waals corrections. According to our findings, the heterostructure possesses appropriate stability and electronic properties for battery applications. The lithium ion is found to strongly trapped at the bridge site (B), at the intersection of the two monolayers, with an intercalation energy of -2.78 eV, much better than previously reported MXene-based heterostructures. The intercalation energy is found to decrease with an increasing Li-ion concentration. Furthermore, these findings revealed that the heterostructure has a small but positive open circuit voltage with a theoretical capacity of 344.02 mAh/g, which makes it a promising anode material for lithium-ion batteries.

Colloidal Dispersion Stabilized By Solvent Structuring: Molecular Simulation Study

Siyoung Q. Choi, Sang Mun Jeong, SeeHoon Lee

This study focuses on developing a framework for understanding the stability of particle dispersions by employing molecular dynamics (MD) and ab initio molecular dynamics (AIMD) simulations. Traditionally, the DLVO theory offers a method to calculate the forces acting between particles, thus predicting their dispersion stability. However, it only works for specific conditions, taking into account electrical repulsive forces, which are influenced by simple factors such as the ionic strength of the solvent and the surface charge of the particles, and a simplified model of van der Waals forces. On the contrast, we show that we can overcome these limitations by simulating the solid/liquid interface, with a particular focus on the interaction between silica (SiO2) and water—a well-known model system for studying dispersion stability. Through simulation, we closely examine the structure of water molecules at the interface and investigates how this structure impacts the dispersion stability of the particles. This approach provides a more comprehensive understanding of the factors that influence the dispersion of particles in a medium, potentially leading to more accurate predictions and control methods for dispersion stability in various applications.

Enhancing Surrogates with Chemical Information

Rohit Goswami

Surrogate construction is one of the key simplifying practical assumptions made in the face of the many body Schrödinger equation. Every attempt at generating a valid surrogate falls somewhere between random sampling with interpolation and the solution of the full system. We will enumerate how to leverage chemical information and statistical knowledge to obtain an optimal sampling method by virtue of constructing multiple surrogates and running saddle searches until the final positional vector coincides with that of the saddle on the ab-initio function of interest. In this manner we show an order of magnitude speed up in terms of the number of calls to the expensive potential and thereby demonstrate scaling of saddle searches for large systems using accurate force fields.

S-Plm: Structure-Aware Protein Language Model Via Contrastive Learning between Sequence and Structure

Qing Shao, Duolin Wang, Jin Chen, Dong Xu, Mahdi Pourmirzaei, Shuai Zeng, Negin Manshour, Farzaneh Esmaili, Yuexu Jiang, Biplab Poudel

Large protein language models (PLMs) present excellent potential to reshape protein research by encoding the amino acid sequences into mathematical and biological meaningful embeddings. However, the lack of crucial 3D structure information in most PLMs restricts the prediction capacity of PLMs in various applications, especially those heavily depending on 3D structures. To address this issue, we introduce S-PLM, a 3D structure-aware PLM utilizing multi-view contrastive learning to align the sequence and 3D structure of a protein in a coordinate space. S-PLM applies Swin-Transformer on AlphaFold-predicted protein structures to embed the structural information and fuses it into sequence-based embedding from ESM2. Additionally, we provide a library of lightweight tuning tools to adapt S-PLM for diverse protein property prediction tasks. Our results demonstrate S-PLM's superior performance over sequence-only PLMs, achieving competitiveness in protein function prediction compared to state-of-the-art methods employing both sequence and structure inputs.

Towards Computation-Guided Design of Tunable Organic-Inorganic CdS Quantum Dot Binary Superlattices

Xin Qi, Chris Lowe, Helen Larson, Jim Pfaendtner, Brandi Cossairt

Inspired by naturally occurring hierarchical materials, numerous beautiful and functional mesoscale architectures have been constructed in labs using solid-state nanoscale building blocks and bio- or biomimetic molecules via self-assembly. Their structures and assembly dynamics are tightly regulated by the system's convoluted energetics and entropic information, as well as morphological variation, which may even respond to a minuscule change. In this study, we observed a CdS quantum dot (QD) superlattice transformation induced by an increment of the peptoid (i.e., the ligand) length by two monomers at a time. However, the complexity of the nanoscale colloidal system often makes it difficult to probe the driving force for lattice transformation via experimental means. Alternatively, we can construct suitable multi-scale theoretical toolkits by carefully selecting and combining methods from different levels to connect the scales and gain physical insights into the assembly mechanisms. Based on theoretical evidence, we show that the assembly is a binary superlattice where both the CdS QD and the peptoid serve as building blocks and further predict a correlation between the superlattice structure and the peptoid sequence. The computationally guided prediction is validated by experiments where superlattice transformation is observed with modified peptoids.

High-Throughput in-Silico Synthesis, Characterization, and Design of Organic Polymer Materials

Timotej Bernat, Michael Shirts, Connor Davel, Jeffrey R. Wagner

Polymer design is a vital part of a wide range of applications including identifying suitable sustainable and recyclable plastics, improving polymer-biopolymer interfaces for biomedical engineering and therapeutics, and designing self-healing materials with dynamic covalent networks. The vast polymer design space over all chemical domains far exceeds the practical amount of experimental capital and research effort available and would greatly benefit from computational structure-function models for systematic exploration. Molecular dynamics (MD) approaches are some of the most attractive of such models due to their speed and interpretability and are becoming sufficiently rapid to evaluate polymer properties on a large scale. However, current open-source MD tools lacks automated, high-throughput, and general infrastructure for constructing and representing polymer systems (so-called '*in-silico* synthesis') which is instrumental in generating structural datasets for bulk physical property evaluation and validation. Here we expand our existing work in polymer modeling to enable high-accuracy polymer design across diverse chemistries and applications using these approaches. The tools we have implemented to these ends have been incorporated into the Open Force Field MD toolkit (https://github.com/openforcefield/openff-toolkit), as well as a purpose-built Python library for general polymer structure generation (https://github.com/timbernat/polymerist), and are described in greater detail in our recent manuscript, published in the ACS Journal of Chemical Information and Modeling (https://pubs.acs.org/doi/10.1021/acs.jcim.3c01691).

Utilizing an Integrated Experimental and In Silico Approach to Engineer Cross Reactive Antibodies Against VEGFA and PlGF2 in Pediatric Glioblastoma

Emma Aldrich, Samuel Blackman, Jennifer Maynard, Kayla Sprenger

Pediatric glioblastoma is characterized by the overexpression of growth factors that target angiogenic pathways in tumors, presenting a unique opportunity for therapeutic design. In healthy cells, Vascular Endothelial Growth Factor-A (VEGFA) contributes to vascular permeability and angiogenesis, but overexpression in cancers is known to contribute to tumor angiogenesis. Meanwhile, Placental Growth Factor 2 (PlGF2) plays a similar role in placental angiogenesis and vascular development, but is also upregulated across many cancer types. Though progress has been made to design antibodies to block VEGFA interactions with its family of Vascular Endothelial Growth Factor Receptors (VEGFRs) and its co-receptor, Neuropilin 1 (NRP1), the introduction of anti-VEGFA antibodies can lead to therapeutic escape by expressing related PlGF2. Critically, sequence identity between growth factors can be leveraged to design cross-reactive antibodies to simultaneously target VEGFA and PlGF2 and their interactions with NRP1 and VEGFR1, presenting opportunities to improve specificity and versatility in vivo and expand clinical translatability. To this end, our experimental collaborators recently determined a suite of novel candidate antibodies and measured their affinity and specificity/cross-reactivity for VEGFA, PlGF2, and antibody-growth factor complexes with NRP1, VEGFR1, and heparin sulfates. This poster will describe our use of computational approaches, including static docking and molecular dynamics simulations, to elucidate the molecular-level interactions between the relevant growth factors and newly-derived antibodies. Additionally, it will highlight our efforts to identify the molecular basis for the apparent cross-reactivity of these antibodies, via assessing the effects of single point mutations introduced into the antibody sequences in silico. Finally, we describe our efforts to quantify competition between the growth factors and antibodies in the presence and absence of NRP1 and VEGFR1. Overall, characterization of this antibody suite informs angiogenic interactions between VEGFA and PlGF2 signaling in pediatric glioblastoma, leveraging in silico approaches to advance efforts toward robust and translatable cancer therapeutics.

Computational and Data-Driven Exploration of the Impact of Metal-Organic Framework Topology on Cryogenic Hydrogen Storage

Kunhuan Liu, Haoyuan Chen, Timur Islamoglu, Xijun Wang, Omar Farha, Randall Snurr

Hydrogen is considered a crucial clean energy vector to mitigate climate change, but due to the low volumetric energy density of gaseous hydrogen, it is difficult to store hydrogen for many practical applications. Cryogenic sorption-based methods, particularly using nanoporous metal–organic frameworks (MOFs), have been considered as viable solutions to enhance the deliverable capacity of stored hydrogen. MOFs are a class of porous crystalline material synthesized from well-defined molecular building blocks (organic ligands and metal clusters) that assemble into a range of networks. The networks denote how building blocks are connected and are often referred as the underlying topology of the MOFs. In this study, we constructed *in silico* 105,764 MOF structures using 534 topologies and performed high-throughput screening of their hydrogen deliverable capacities using grand canonical Monte Carlo (GCMC) simulations and surrogate machine learning models. Based on the analysis of over 100,000 MOFs, we explored the less-known effect of underlying topologies of MOFs on hydrogen deliverable capacity. We discover that while most topologies can generate high-performing MOFs, the best performing MOFs are generated with a set of topologies that have a low average deliverable capacity. We analyzed mathematical descriptors of the topologies including net density and td10 to understand the correlation between topologies and textural properties as well as H_2 capacities. We subsequently apply these principles to identify promising topologies that are experimentally feasible.

Hydrogen is considered a crucial clean energy vector to mitigate climate change, but due to the low volumetric energy density of gaseous hydrogen, it is difficult to store hydrogen for many practical applications. Cryogenic sorption-based methods, particularly using nanoporous metal–organic frameworks (MOFs), have been considered as viable solutions to enhance the deliverable capacity of stored hydrogen. MOFs are a class of porous crystalline material synthesized from well-defined molecular building blocks (organic ligands and metal clusters) that assemble into a range of networks. The networks denote how building blocks are connected and are often referred as the underlying topology of the MOFs. In this study, we constructed *in silico* 105,764 MOF structures using 534 topologies and performed high-throughput screening of their hydrogen deliverable capacities using grand canonical Monte Carlo (GCMC) simulations and surrogate machine learning models. Based on the analysis of over 100,000 MOFs, we explored the less-known effect of underlying topologies of MOFs on hydrogen deliverable capacity. We discover that while most topologies can generate high-performing MOFs, the best performing MOFs are generated with a set of topologies that have a low average deliverable capacity. We analyzed mathematical descriptors of the topologies including net density and td10 to understand the correlation between topologies and textural properties as well as H_2 capacities. We subsequently apply these principles to identify promising topologies that are experimentally feasible.

Formally-Verified Energy Calculations for Benchmarking Molecular Simulations.

Ejike Ugwuanyi, Tyler R. Josephson, John Velkey, Colin Jones

This study introduces an approach to computing the Lennard-Jones potential for molecular simulations using Lean 4. Lean 4 is both an interactive theorem prover and a purely functional programming language that allows writing and proving theorem statements about executable functions. By connecting formal proofs with efficient programming, we pave the way for computational rigor rarely seen in traditional molecular simulations. The focus on formalization and accuracy is essential for theoretical chemistry, where precise simulations can lead to a better understanding of chemical phenomena [1].

We implement periodic boundary conditions (PBC) and the minimum image convention (MIC) for a system of Lennard-Jones particles, and we also write formal proofs about these concepts. Common languages like Python and C use for and while loops, imperative programming techniques that are absent in Lean, a purely functional programming language. In this study, we first translate the loop-based energy calculation into recursive functions in Python, ensure benchmark against the NIST benchmarks [2], then translate the functions from Python into Lean 4. These functions can be the subject of formal proofs that verify their mathematical and logical properties. The Lean implementation is a stronger benchmark than that available by NIST because the underlying mathematics have been formally verified.

In the future, we aim to implement and formalize the velocity Verlet algorithm and neighbor lists, expanding the scope beyond the energy calculation to enable formally verified molecular dynamics simulations. This approach elevates the standards of computational verification in chemistry and promises broader applications in science.

[1] Bobbin, Maxwell P, et al., “Formalizing Chemical Physics Using the Lean Theorem Prover,” Digital Discovery, 3, 16 Nov 2024, p 264–280, https://doi.org/10.1039/D3DD00077J.

[2] https://www.nist.gov/mml/csd/chemical-informatics-group/lennard-jones-fluid-reference-calculations-cuboid-cell

Characterizing the Suitability of Metal Organic Polyhedra for Gas Sorption

Omkar Desai, Andrew J. Schultz, David Kofke

Metal-organic polyhedra (MOP) are made up of organic and metallic structural building units. In this respect, they have features in common with metal-organic frameworks (MOF). Unlike MOFs, which use the metal+ligand moieties to build an extended lattice framework, MOPs are constructed as discrete, finite units, typically shaped as some type of polyhedron, such as a cube or bipyramid; this has led MOPs to be described as "the molecular equivalent of MOFs". These structures offer tailored porosity and functionality, and high surface area, making them ideal candidates for membrane fabrication, particularly as part of a mixed-matrix membrane material. MOP can be used for gas storage, catalysis, and various other applications. Metallic and organic sub-units affect gas-MOP interaction. As with MOFs, MOPs present a huge variety of design choices to promote a desired functionality. Molecular modeling can help guide these choices.

Comparison of MOP models to experimental data is highly problematic, and cannot be performed in a definitive, quantitative manner. Experimental MOP data exists only for MOPs either in an extended lattice, or dispersed in polymer. Performing and interpreting simulations of such complex systems is fraught, so we seek a way to characterize the affinity of a solute to the MOP in isolation. Affinities in extended frameworks, such as MOFs, are normally characterized by the Henry's constant. This can be computed from a molecular model using methods such as Widom insertion. To quantify adsorption affinity to an isolated structure such as a MOP, a metric based on the Henry's constant does not quite fit the situation where there is no well-defined simulation volume. Instead, we examine a metric that isolates the MOP-solute interaction without requiring a volume specification. The quantity is akin to a virial coefficient, and can be computing using established techniques, such as Mayer-sampling Monte Carlo. We examine this metric for several MOP-solute systems, using the UFF model, and we compare it qualitatively and quantitatively to available experimental sorption data for these systems.

Can Classical Nucleation Theory Describe Heterogeneous Crystal Nucleation on Non-Uniform Surfaces?

Fernanda Vargas, Sarwar Hussain, Amir Haji-Akbari

Crystal nucleation stands as a pivotal phenomenon in numerous natural and industrial processes. Its profound implications extend to high-end industries, including semiconductors, solar cells, and aviation, where the understanding alongside the manipulation of heterogeneous crystal nucleation is of utmost importance as it is at the heart of the production of single crystalline materials. The time and length scales at which nucleation proceeds are too small to capture using most experimental techniques, presenting a challenge to experimental investigations of nucleation. As such, computer simulations are utilized to unveil the molecular details of the nucleation process. Historically, classical nucleation theory (CNT) has been used extensively to describe the physics of both homogeneous nucleation and heterogeneous nucleation. Despite its success, its validity or lack thereof under different circumstances is still debated.

In this work, we use molecular simulations and advanced path sampling techniques to probe heterogeneous nucleation of the Lennard-Jones (LJ) crystal on different uniform and patterned surfaces. To rigorously assess the applicability of CNT, we first compute heterogeneous nucleation rates using the jumpy forward flux sampling algorithm and examine their dependence on temperature. We also assess the stability of the FCC and HCP crystals next to each surface by conducting conventional MD simulations of FCC and HCP crystals at its vicinity and characterizing the extent of pre-melting. Finally, we also analyze crystalline nuclei along the nucleation pathway and find them to change shape upon enlargement, directly contradicting one of the core assumptions of CNT that crystalline nuclei are spherical caps with fixed contact angles. Overall, our findings underscore the nuances and intricacies of applying CNT to probe the kinetics of heterogeneous nucleation.

Protein Unfolding Behavior Under Varying External Stressors: Insights from Molecular Dynamics Simulations

Yinhao Jia, Janani Sampath

The field of protein therapeutics has undergone rapid development during the last decade, as more and more biologics are being approved by the FDA every year. However, the inherent instability of proteins under non-native conditions—such as variations in temperature and pH encountered during production, transportation, and administration—poses a significant challenge to their therapeutic applications. Their efficacy can be affected dramatically due to the loss of their native structure. The mechanisms underlying this degradation vary, as different stressors may affect distinct aspects of a protein structure. To this end, knowledge about a protein's weak points is of great importance in the strategic design of new materials, such as polymers, to protect ephemeral protein therapeutics and improve their stability when subjected to a loss of function induced by external stressors.

In this work, we perform all-atom molecular dynamics simulations to elucidate the unfolding mechanisms of proteins subjected to external stressors, including temperature, pH, and shear stress. We analyzed proteins of varying sizes, such as lysozyme (~14 kDa), Bovine Serum Albumin (~67 kDa), and CRISPR-BrCas12b (~130 kDa), using general structural analysis like root-mean-square deviation (RMSD), hydrogen bond number, and solvent accessible surface area (SASA) to capture the onset of unfolding. While indicative of structural destabilization, these metrics offer limited insight into the specific unfolding pathways. An in-depth investigation of the secondary structure evolution helps explain the unfolding process, which may complicate the process if a larger protein, like BSA, is analyzed. Separation of the protein structure analysis into defined functional domains provides a better interplay of interaction between domains and gives insight into the unfolding process. Utilizing these metrics can distinguish the unfolding pathway triggered by varying stressors. Other analyses, including fractions of native contact and residue contact analysis, offer a comprehensive view of the unfolding process and the potential weak connections. We can use this information to rationally design polymer conjugates to target the weak point and shield proteins from specific stressors.

Effect of Pressure on the Conformational Landscape of Human Γd-Crystallin from Replica Exchange Molecular Dynamics Simulations

Arlind Kacirani, Betul Uralcan, Tiago Domingues, Amir Haji-Akbari

Human γD-crystallin belongs to a crucial family of proteins known as crystallins located in fiber cells of the human lens. Since crystallins do not undergo any turnover after birth, they need to possess remarkable thermodynamic stability. However, their sporadic misfolding and aggregation, triggered by environmental perturbations or genetic mutations, constitute the molecular basis of cataracts, which is the primary cause of blindness in the globe according to the World Health Organization. Here, we investigate the impact of high pressure on the conformational landscape of the wild-type HγD-crystallin using replica exchange molecular dynamics simulations augmented with principal component analysis. We find pressure to have a modest impact on global measures of protein stability, such as root mean square displacement and radius of gyration. Upon projecting our trajectories along the first two principal components from PCA, however, we observe the emergence of distinct free energy basins at high pressures. By screening local order parameters previously shown or hypothesized as markers of HγD-crystallin stability, we establish correlations between a tyrosine-tyrosine aromatic contact within the N-terminal domain and the protein's end-to-end distance with projections along the first and second principal components, respectively. Furthermore, we observe the simultaneous contraction of the hydrophobic core and its intrusion by water molecules. This exploration sheds light on the intricate responses of HγD-crystallin to elevated pressures, offering insights into potential mechanisms underlying its stability and susceptibility to environmental perturbations, crucial for understanding cataract formation.

Navigating Polymer Membrane Design: Balancing Durability and Performance Using Molecular Insights

Mohammed Al Otmi, Ping Lin, Diana Aponte Claudio, Coray Colina, Janani Sampath

Polymer membranes are crucial in numerous industrial applications, including gas separation and ion exchange processes. Polymers offer a broad design space, and their properties can be probed using pre- or post-synthetic approaches. Glassy polymers, known for their high free volume, have found use in many separation applications due to their high permeability and selectivity. Free volume elements refer to the void distribution in the polymer matrix and play a significant role in the performance and durability of the membrane. Transport, in general, is governed by molecular-level physics that are challenging to characterize using experimental approaches. Molecular dynamics simulations allow us to examine short length and timescales that enable us to capture and understand these physics. Polymer relaxation due to plasticization or aging is a major hurdle in both gas separation membranes and anion exchange membranes. Effectively modeling the transport and unraveling the relaxations dynamics is essential for enhancing membrane performance and durability.

In this study, we investigate the distribution of free volume elements (FVE) in gas separation and anion exchange membranes. Gas transport within these membranes is understood as a series of thermally activated jumps between voids distributed throughout the polymer matrix. These FVEs have a transient nature and their dynamics play a pivotal role in determining membrane transport properties. Our analysis reveals a clear correlation between FVE structure and backbone rigidity, as well as orientational and translational segmental dynamics, particularly in the FVE interface. This correlation is instrumental in characterizing phenomena such as plasticization (swelling) and aging (densification), facilitating the design of membranes that exhibit resistance against these phenomena. Optimizing anion exchange membranes (AEMs) for alkaline fuel cells presents an ongoing research challenge in fuel cell applications. AEMs often suffer from low conductivity and a susceptibility for high water swelling, this swelling compromise their mechanical integrity. We employ detailed FVE analysis to identify hydration channels which are crucial in facilitating effective vehicular transport across the membrane. Our examination of various polyolefin-based AEMs reveals a notable association between interconnected water channels and enhanced hydroxide (OH^-) transport. Through extensive simulations, we calculate hydroxide diffusion across different chemistries and analyze mechanical performance by subjecting membranes to uniaxial deformation. Our simulations capture the tradeoff between hydroxide transport and mechanical integrity within the membrane module.

Future work focuses on quantifying the contribution of specific molecular motions to membrane flexibility through enhanced sampling approaches such as metadynamics. By employing these calculations, we can rationally design the membrane backbone, enabling us to manipulate chemistries to enhance the stability of the overall structure. Unraveling the molecular origins of relaxation processes—such as swelling and aging—is imperative for enhancing membrane durability and sustaining high performance over long lifetimes.

Controlling Reaction Microenvironment in Electrochemical CO_2 Reduction through Polymer-Ligated Metal Nanoparticles

Joseph Tapia, Qiang Luo, LE Zhou, Chung-Hao Liu, Maham Liaqat, Hanyi Duan, Zhefei Yang, Mu-Ping Nieh, Todd Emrick, Jie He, Peng Bai

Metal nanoparticles (NPs) exhibit unique catalytic activities but are prone to sintering. It has recently been demonstrated that functionalizing the NP surface with polymer ligands can inhibit sintering and, in addition, lead to improved selectivities for CO2 reduction over the hydrogen evolution reaction. To understand the impact of surface-attached polymers, we performed molecular dynamics simulations to compare three different polymers, including hydrophobic polystyrene (PS), hydrophilic poly(ethylene oxide) (PEO), and zwitterionic poly(1H,1H,2H,2H-perfluoro-1-octanol-2-methacryloyloxyethyl choline phosphate (poly(PFO-MCP)). Polymer chains of 44 – 65 repeating units were grafted to a gold slab with exposed (111) surface and immersed in an aqueous environment with 2000 – 3000 water molecules. We found that both hydrophobic and zwitterionic ligands increase the relative concentration of CO_2 near the NP surface as CO_2 preferentially resides in the polymer region rather than in the bulk water region. The choice of ligands shows great effect on water structuring on the surface above the NP. PS was found to generate highly structured water droplets throughout the polymer region while poly(PFO-MCP) induces extended chains of diffuse water molecules due to the amphipathic nature of the monomer repeating unit containing zwitterionic heads and fluorinated tails. This difference in solvent structuring leads to an interesting difference in CO_2 diffusion. We found that the effective diffusivity of CO2 in poly(PFO-MCP) along the surface normal is four times greater than the overall 3-dimensional diffusivity, whereas no such anisotropy was observed in PS. The results of this study offer insights into how surface modification using polymer ligands can help create a microenvironment to drive reactivity of metal NPs.

Effect of Interface Curvature at the Axial Junction in Silicon-Germanium (Si-Ge) Nanowires

Oreoluwa Adesina

Thermal management, thin-film technologies, microelectronics, and semiconductor and optical devices, as well as low-dimensional thermoelectric devices, are examples of the wide array of potential applications that silicon and germanium can have. Thanks to these applications, thermal (heat) transport in silicon-germanium (Si-Ge) materials, including those at the nanoscale – wires, is an area of continued interest. The interfacial thermal resistance governs thermal transport in small-scale heterogeneous systems but the relative contributions of different transport mechanisms (e.g., elastic and inelastic phonon scattering) remain elusive. The curvature and sharpness of the interface at the heterojunctions in Si-Ge nanowires depend on their growth environment. In our research, we assess the effect of interfacial curvature at the axial junctions in Si-Ge nanowires. We employ the non-equilibrium molecular dynamics (NEMD) Müller-Plathe method (with the Tersoff potential) to model thermal transport in nanowires with lengths ranging between 100 nm and 500 nm and that possess varying interfacial curvatures. The resistance at the interfaces of the heterojunctions is estimated. With different radii sizes controlling the curvature shapes at the Si-Ge interface, models are simulated, and the boundary resistance is probed for directional transport between first, Si and Ge, and second, Ge and Si.

Cross-Linked Covalent Organic Frameworks for Maintaining Layer Order in Two-Dimensional Membranes

Alathea Davies, Michael Wenzel, Bruce Parkinson, John Hoberg, Laura de Sousa Oliveira

A primary challenge for implementing two-dimensional materials into membranes is that membrane wetting often leads to flake dispersion. This can increase the tortuosity of permeating particles and reduces their permeance rate. Cross-linking, or "chemically stitching", a material together can help maintain the well-ordered layers and reduce the interstitial flow of particles. While this is not a new concept for polymer chemistry, it has little study in covalent organic frameworks (COFs). This combined computational and experimental study investigates the feasibility of cross-linking a quinoxaline-based COF for liquid separation applications. Self-consistent charge density functional tight-binding (SCC-DFTB) was used to model the COF with two different linking moieties: oxalyl chloride (OC) and hexafluoroglutaryl chloride (HFG). Equations were derived for cohesive energy and enthalpy of formation to quantitate the stability and ease of formation, respectively. The cross-linked COFs had higher cohesive energies than the non-cross-linked COF, indicating that the intramolecular bonding of the COF was stabilized by the cross-linking reagents. Enthalpy of formation values also supported thermodynamic feasibility. In combination with semi-empirical molecular dynamics, it was determined that the HFG moiety resulted in more stable COFs. The HFG cross-linked COF was synthesized and characterized using Fourier transform infrared spectroscopy (FTIR), X-ray diffraction (XRD), thermogravimetric analysis (TGA), and water contact angle measurements, all confirming the presence of HFG cross-linking. Liquid permeance testing and TGA confirmed that a saturation limit for cross-linking had been experimentally reached, and while that saturation was not as high as the models can achieve, additional computational analysis on cross-linking concentrations indicates that decreased cross-linking has little effect on overall stability.

Analysis of RAST Activity Coefficients for Ethanol/Water Adsorption in All-Silica Zeolites

Anne Le, Yachan Liu, Peng Bai

Nanoporous zeolites are materials widely used for adsorption- or membrane-based separation processes, but there are key gaps in knowledge when examining adsorption behavior in complex yet realistic systems. In practice, the Ideal Adsorbed Solution Theory (IAST) of Myers and Prausnitiz is often used to predict mixture isotherms from pure-component data, even though significant deviations from IAST have been reported for many industrially relevant, non-ideal mixtures. Despite the importance of mixture adsorption knowledge, directly measuring these non-ideal isotherms is a challenge both experimentally and computationally. Leveraging the simplicity of measuring single-component isotherms and the wide availability of activity coefficients in non-confining systems, this work aims to relate zeolite structure to pore-phase activity coefficients as a general strategy for modeling mixture adsorption behavior. As the first step towards this goal, we extend our previous study on the adsorption of binary water/ethanol mixtures in zeolites and apply the Real Adsorbed Solution Theory (RAST) to extract pore-phase activity coefficients in hundreds of zeolites catalogued by the International Zeolite Association. Grand-canonical Monte Carlo simulations were used to calculate pure and binary isotherms, and RAST activity coefficients were modeled using the Margules activity model. Comparing ethanol-selective zeolites varying in pore size, we observe activity coefficients can be framework dependent. For example, medium- and large-pore zeolites (e.g., MFI, BEA, FAU) generally have activity coefficients larger than those in bulk solution, while activities in small-pore zeolites (e.g., LTA zeolite) are smaller than bulk. We examine how these observations generalize across different zeolites and anticipate these results will provide new insights into how confinement modifies bulk nonidealities.

Minimization of Adsorption Hysteresis: Making Adsorption Reversible

Filip Formalik, Jesse Prelesnik, Joern Siepmann, Randall Snurr

Adsorption hysteresis can significantly impact the functionality of porous materials like metal-organic frameworks (MOFs), influencing the reversibility of adsorption in applications such as adsorption cooling and water harvesting, where reversibility is essential. Hysteresis is especially pronounced in MOFs with large pores, which hinders the use of these materials despite their potential for high adsorption capacity. Adsorption hysteresis arises from the presence of high energy barriers separating low- and high-density phases of the fluid confined in the pore and is coupled with the tendency of the system to remain in metastable phases.

In this study, we explore adsorption hysteresis through the van der Waals (vdW) loop, a metric that quantifies the stability limits of adsorption hysteresis. The vdW loop maps out the density of stable, metastable, and unstable phases of the confined fluid, utilizing spinodal and binodal points to define the stability limits of the phases and the equilibrium pressure for coexisting phases, respectively. To accurately predict the vdW loop, we employ transition matrix Monte Carlo (TMMC) simulations. These simulations offer a comprehensive description of the free energy landscape associated with adsorption, enabling the capture of all possible system macrostates, including unstable regions of the phase diagram (e.g., barriers between states). Our analysis leverages the gRASPA simulation software, developed in our group and optimized for GPU nodes, to significantly accelerate the adsorption simulations.

Our findings illustrate the effects of the simulation system's size on the width of the hysteresis loop. These effects, in turn, correlate with the barrier associated with the transition between low- and high-density phases. We further explore how specific interaction patterns on the surfaces of idealized pore geometries (such as *hydrophobic* and *hydrophilic* sites) can influence these properties. Patterns explored in model systems can be correlated with the distribution of functional groups in real MOFs. This relationship is important for the design of MOFs optimized for minimal adsorption hysteresis and for enhancing their efficiency in applications like adsorption cooling and water harvesting. Additionally, we establish connections between the hysteresis loop, the transition barrier, and particular pore shapes and system sizes in MOFs, offering new insights into the design of new materials showing reversible adsorption.

Understanding Dissolution of $MAPbI_3$ Perovskites in Water with Machine Learning Derived Interatomic Potential

Nisarg Joshi, Nadia Intan, Marlo Zorman, Jim Pfaendtner

The degradation of hybrid organic inorganic perovskites (HOIP) in liquid water has limited its application to photovoltaics[1] and clean energy methods. There have been several studies which use first principle calculations[2,3] to understand this degradation mechanism. However, all the studies have reported different mechanisms and the primary mechanism for $MAPbI_3$ dissolution remains in dispute.[4] In addition, the study of degradation mechanisms is limited by the computational expense of first principle methods. In this study, we propose to understand the dissolution of $MAPbI_3$ with machine learning potentials (MLP). Trained on accurate ab initio molecular dynamics (AIMD) data, we explore degradation pathways of perovskites with the high accuracy of its training data and a small fraction of the computational cost. We enhance the feasibility of MLP to examine degradation pathways by developing training protocols which require less AIMD data. We also illustrate the generalizability of ML potentials for different surfaces and explore the extent to which they can be used for multiple surfaces and provide analysis into the mechanism.

References:

1. Massimo Spina, Ayat Karimi et al. Mechanical signatures of degradation of the photovoltaic perovskite CH3NH3PbI3 upon water vapor exposure *Appl. Phys. Lett.* 110, 121903 (2017)
2. Chao Peng, Jianfu Chen, Haifeng Wang, and P. Hu The Journal of Physical Chemistry C 2018 *122* (48), 27340-27349 DOI: 10.1021/acs.jpcc.8b07294
3. Chao Zheng and Oleg Rubel The Journal of Physical Chemistry C 2019 *123* (32), 19385-19394 DOI: 10.1021/acs.jpcc.9b05516
4. Nadia Intan and Jim Pfaendtner ACS Nano 2023, 17, 22371-22387

Effect of Electrode-Electrolyte Interface on C—C Coupling for C_{2+} Product Formation on $PdCu_3$

Jayendran Iyer, Tuhin Suvra Khan, Debra Bernhardt, M. Ali Haider

The C—C coupling step is a crucial elementary step in CO_2 reduction reaction (CO_2RR) to estimate the selectivity of C_{2+} products over transition metal catalysts. However, the mechanism of this elementary step is debatable due to highly unstable OCCO* intermediate proposed earlier. We, in this work, show the C—C coupling as a concerted mechanism with proton coupled electron transfer as the key route with experimental evidence. Here, we analyze the role of Pd and Cu in (100) and (111) facets of $PdCu_3$ intermetallic alloy. In sync with the experimental observation, from our ab initio calculations, we can confirm (100) facets to be more active than (111) facets. In addition, the C—C coupling barrier with Pd atoms at the surface on $PdCu_3$ (100) facets is observed 60 kJ/mol higher than in the absence of surface Pd atoms resulting in overall increase in C_{2+} products formation rate. This is attributed to the d-band center shift observed away from the fermi level in the absence of surface Pd atoms. However, no d-band center shift is evident with different Pd configurations on $PdCu_3$ (111) surface leading to similar C—C coupling barriers of ~90 kJ/mol over these surfaces. On investigating further into the effect of electrode-electrolyte interface over these $PdCu_3$ surfaces, we observe a clear influence of K^+ ion and solvent molecules on the C—C coupling barrier. With the increase in intermediate coverage, the C—C coupling barrier drops from 200 kJ/mol to 50 kJ/mol. This shows a huge influence of lateral interactions between the surface intermediates resulting in improved C_{2+} products. Moreover, the distance of K^+ and the transition state structure also helps in reducing the coupling barrier.

Reaction-Aware Machine Learning Potentials Using Transition-State Finding Methods

Siddarth Achar, Priyanka Bholanath Shukla, Caitlyn Vinger, Chinmay Mhatre, Leonardo Bernasconi, Karl Johnson

Machine learning forcefields are powerful because they can map molecular structures to high-dimensional potential energy surfaces with near-quantum accuracy, but with a fraction of the computational cost of quantum calculations. An advantage of most machine learning forcefields is the use of molecular descriptors, which enable encoding of information about chemical reactions without the need to explicitly add bonding terms, unlike most empirical forcefields. However, training machine learning forcefields to accurately account for a wide range of chemical reactions can be extremely inefficient. Therefore, there is a critical need to develop algorithms to optimize the exploration of configurations that are essential to the reactive part of the potential energy surface. Most common active learning methods for developing forcefields for specific chemical reactions often require long molecular dynamics simulations to capture reactive events. Alternatively, one could use enhanced sampling methods with a predefined reaction coordinate to explore different configurations. However, this introduces human bias in choosing the reaction coordinates or the collective variables. We propose a method that tackles this limitation by automatic generation of potential reactions, coupled with transition-state (TS) finding methods that can generate training data without specifying reaction pathways or desired products. We employed the DeePMD formalism to train our forcefields in an iterative active learning scheme. These inexpensive forcefields were used to explore reactive space via automatic generation of potential reactions and TS finding codes like the single-ended growing string method and the nudged elastic band method. This active learning scheme was tested for two gas phase reactions: ammonia synthesis and methylene imine hydrolysis, and a heterogenous catalytic reaction: methane coupling on titanium carbide. We observed accurate predictions of reaction barriers for these reactions via our active learning scheme. We found that our method reduced the total number of configurations needed to achieve the same performance as other active learning methods.

An Integrated Off-Lattice Kinetic Monte Carlo (KMC)-Molecular Dynamics (MD) Framework for Modeling Polyvinyl Chloride Dehydrochlorination

Feranmi Olowookere, Heath Turner

Many plastic wastes, including polyvinyl chloride (PVC), contain considerable repositories of molecular value. These resources can potentially be accessed via depolymerization mechanisms; we aspire to perform these transformations using scalable techniques at mild conditions. In this study, a three-dimensional off-lattice kinetic Monte Carlo-molecular dynamics (KMC-MD) framework [*Comp. Mat. Sci.* 229, 112421 (2023)] is developed to investigate the dehydrochlorination/conjugation transformation of PVC in sodium hydroxide (NaOH) with atomistic resolutions at experimental timescales ($10^3 - 10^6$ s). Our framework enables a comprehensive examination of the competing reaction pathways and molecular-scale changes influenced by various solvents (acetone, ethylene glycol, triethylene glycol, tetrahydrofuran, and bio-derived solvents), as well as the influence of varying molecular weight distributions, concentrations, and temperatures. The algorithm simulates bond cleavage and formation during the KMC stages, whereas the MD stage is dedicated to the relaxation and thermalization of the PVC-NaOH-solvent system. The framework allows us to capture important configurational aspects (mixing, correlations, clustering, etc.) that are not accessible with a traditional microkinetic model. Our KMC-MD approach provides an exceptionally powerful tool for elucidating the complex reaction pathways, while simultaneously gaining insights into the molecular changes inherent to the depolymerization processes. Our technique allows us to perform direct experimental benchmarking at experimental timescales, paving the way for more effective upcycling of polymer wastes into valuable chemicals under mild conditions.

Examination of the Virial Series As a Molecular-Based Equation of State for Gas and Supercritical-Fluid Mixtures

Raghavendran Suresh, Omkar Desai, Andrew J. Schultz, David A. Kofke

The virial equation of state (VEOS) is unique in its ability to provide a rigorous thermodynamic equation of state from a model of the intermolecular potential-energy surface (PES). Virial coefficients B_n that enter into the VEOS are specified as integrals of the PES for 2, 3, etc. molecules. The PES may be based on semi-empirical force fields, or developed from ab initio quantum chemical calculations that are increasingly being employed as an input to machine-learning based PES representations. In the latter application, the VEOS then provides a thermodynamic model that is built completely on first principles. In either case, knowledge of the VEOS from a PES provides a critical tool for evaluating the quality of the PES, because the properties computed from the VEOS (including some of the coefficients themselves) can be compared directly to experiment. This provides a precise determination of the quality of two-body, three-body, etc. components of the PES, presenting a rigorous assessment of its accuracy while guiding improvements.

In this study, we investigate the effectiveness of the VEOS as a molecular-based equation of state by computing pure and mixture gas-phase virial coefficients for 30 compounds, including common alkanes, alcohols, benzene-derivatives and small-molecule gases. In contrast to coefficients computed from a first-principles PES, which should in principle match virial coefficients from experiment directly, coefficients computed from pairwise-additive semi-empirical force fields are expected to exhibit a cancellation of errors that lead them to provide values of B_2, B_3, etc. that are inaccurate, but in an offsetting manner. Therefore, we focus comparison of the VEOS derived from these effective potentials to PVT property data, rather than the coefficients themselves. For our initial study, we have chosen Universal Force Field (UFF) and OPLS-AA force fields in our calculations due to their broad applicability to a wide range of compounds. We calculate gas-phase pure and mixed virial coefficients up to the fourth virial coefficient B_4 over a range of temperatures from 100 K to 1000 K using Mayer-sampling Monte Carlo with application of the correction for flexibility as necessary. We have also computed temperature derivatives of these virial coefficients to enable calculation of thermal properties such as the Joule-Thomson coefficient and heat capacities. We validated our calculated virial coefficients with experimental PVT-data provided by the NIST ThermoData Engine, showing convergence of the pressure-density isotherms constructed with the VEOS truncated up to various orders. The comparison points to strengths and weaknesses of these force-field models. We expect that this work would bring more awareness to the effectiveness of the VEOS as an alternative property prediction tool to traditional molecular simulation, and its value in validating and guiding development of molecular models.

Considerations in the Use of ML Interaction Potentials for Free Energy Calculations

Orlando A. Mendible Barreto, Jonathan K. Whitmer, Yamil Colón

Machine learning potentials (MLPs) offer the potential to accurately model the energy and free energy landscapes of molecules with the precision of quantum mechanics and an efficiency similar to classical simulations. This research focuses on using equivariant graph neural networks MLPs due to their proven effectiveness in modeling equilibrium molecular trajectories. A key issue addressed is the capability of MLPs to accurately predict free energies and transition states by considering both the energy and the diversity of molecular configurations. We examined how the distribution of collective variables (CVs) in the training data affects MLP accuracy in determining the free energy surface (FES) of systems, using Metadynamics simulations for butane and alanine dipeptide (ADP). The study involved training forty-three MLPs, half based on classical molecular dynamics data and the rest on ab initio computed energies. The MLPs were trained using different distributions that aim to replicate hypothetical scenarios of sampled CVs obtained if the underlying FES of the system was unknown. Findings for butane revealed that training data coverage of key FES regions ensures model accuracy regardless of CV distribution. However, missing significant FES regions led to correct potential energy predictions but failed free energy reconstruction. For ADP, models trained on classical dynamics data were notably less accurate, while ab initio-based MLPs predicted potential energy well but faltered on free energy predictions. These results emphasize the challenge of assembling an all-encompassing training set for accurate FES prediction and highlight the importance of understanding the FES in preparing training data. The study points out the limitations of MLPs in free energy calculations, stressing the need for comprehensive data that encompasses the system's full FES for effective model training.

Structure-Property Relationships for the Electrothermic and Thermophysical Properties of Ionic Liquids Under External Electric Fields

Fernando Carmona Esteva, Yong Zhang, Edward Maginn, Yamil Colón

New advancements in chemical separation processes are necessary to sustainably meet the increasing demand for electronics.[1] Particularly, progress is needed in solvent design under the influence of external stimuli.[2] Ionic liquids (ILs) offer an opportunity to investigate the effect of external stimuli such as external electric fields (EEFs) on separation media. ILs have thermophysical properties that make them attractive solvents for sustainable separation processes such as low volatility, high thermal stability, and tuneability of their thermophysical properties.[3] Moreover, their charged nature makes them a great test case to investigate how the direct effects of EEFs on nanoscale dynamics and structure can impact the thermophysical properties of solvents under external stimuli. We uncover structure-property relationships for the thermophysical properties of ILs under EEFs by exploring the effects of external stimuli on structure, dynamics, and thermophysical properties through molecular dynamics simulations. For effects on structure, we quantify the effect of EEFs on nanoscale organization through excess entropy and its immediate consequences to dynamics by testing scaling relationships between self-diffusion coefficients and excess entropy under EEFs.[4] Through excess entropy relationships, we are able to describe the entire dynamical landscape of a mixture and correlate systems that showcase distinct reactions to dynamics under external stimuli. With this understanding of dynamics, we also elucidate the emergent electrothermic properties of charged solvents in the presence of EEFs such as the electrostrictive and electrocaloric coefficients. Finally, with the implementation of novel methodologies to detect glass transition temperature from MD simulations, we explore the emergent phase-changing effect of EEF on ILs by correlating changes in excess entropy in the liquid phase to changes to the glass transition temperature.[5] Such analysis allows us to establish connections between the glass and liquid phases and showcases how MD simulations can be leveraged to investigate the emergent behavior of ILs under EEFs, gain insights on the effects of EEF on the structural organization, and establish a better understanding of their thermophysical properties. With this insight we construct structure-property relationships for transport and thermophysical properties of IL in the presence of an EEF. Unlocking the control of transport and thermophysical properties through the manipulation of IL structure and EEF and eventually allowing for the next paradigm of separation schemes.

References:

(1) Eggert, R.; Wadia, C.; Anderson, C.; Bauer, D.; Fields, F.; Meinert, L.; Taylor, P. Rare Earths: Market Disruption, Innovation, and Global Supply Chains. *Annu Rev Environ Resour* **2016**, *41* (1), 199–222. https://doi.org/10.1146/annurev-environ-110615-085700.

(2) *A Research Agenda for Transforming Separation Science*; National Academies Press: Washington, D.C., 2019. https://doi.org/10.17226/25421.

(3) Welton, T. Ionic Liquids: A Brief History. *Biophys Rev* **2018**, *10* (3), 691–706. https://doi.org/10.1007/S12551-018-0419-2.

(4) Dyre, J. C. Perspective: Excess-Entropy Scaling. *J Chem Phys* **2018**, *149* (21). https://doi.org/10.1063/1.5055064.

(5) Carmona Esteva, F. J.; Zhang, Y.; Colón, Y. J.; Maginn, E. J. Molecular Dynamics Simulation of the Influence of External Electric Fields on the Glass Transition Temperature of the Ionic Liquid 1-Ethyl-3-Methylimidazolium Bis(Trifluoromethylsulfonyl)Imide. *J Phys Chem B* **2023**, *127* (20), 4623–4632. https://doi.org/10.1021/acs.jpcb.3c00936.

Unraveling Lasso Peptide Folding: Tapping the Synergy between Machine Learning and Chemical Intuition

Gabriel Costa Alverni da Hora, Jessica M. J. Swanson

Lasso peptides, a distinct class of natural products, are characterized by their unique knot-like tertiary structures. These structures confer extraordinary stability to lasso peptides, making them resistant to proteolytic degradation, thermal denaturation, and extreme pH conditions. Despite these promising characteristics, the direct synthesis of lasso peptides remains a formidable challenge due to the rarity of spontaneous lasso folding.[1] This work introduces a novel approach that merges chemical intuition and machine learning to identify collective variables (CVs) for the relative folding propensity of lasso peptides.

Our research started centered on microcin J25 (MccJ25), a lasso peptide known for its antimicrobial properties. We have engineered new algorithms to distinguish threaded from non-threaded precursors and determine handedness. Our findings reveal that MccJ25 forms right-handed pre-lassos, consistent with all-natural lasso peptides. Furthermore, we demonstrate that strategic *de novo design* of the ring-forming residues with organic moieties and the His5 protonation increase pre-lasso ensemble stability.[2] The use of machine learning employing Harmonic Linear Discriminant Analysis (HLDA) has allowed us to discover CVs capable of distinguishing the pre-lasso fold and converging the folding propensity in well-tempered metadynamics simulations. While the pure chemical-intuited CV has failed, the combination of ring-tail Cα distances and the Gly1N—Glu8δ isopeptide bond distance emerged as effective discriminators of the pre-lasso and pre-tadpole ensembles. Lastly, we tested these CVs on two other lasso sequences, Ulleungdin and Sungsanpin. The results were encouraging, with CVs (the combination of machine learning and chemically intuited) proving to be transferable. This suggests the potential of a universal protocol for studying diverse lasso peptide sequences. Our findings underscore the critical role of stabilizing ring-tail interactions in the quest to fold the elusive pre-lasso motif and highlight the potential of these CVs as a powerful tool for the computational prediction and design of lasso peptides.[3]

This work represents a substantial stride toward the development of a generalizable platform for the chemical synthesis of lasso peptides. The insights gained from this study could pave the way for the design of innovative therapeutics based on lasso peptides. The amalgamation of intuitive CVs and machine learning approaches provides a new perspective in the study of lasso peptides and their intricate folding pathways.

1. Ferguson, A. et al. *Biophys. J.* **2010**, 99 (9), 3056.
2. da Hora, G. C. et al. *J. Am. Chem. Soc.* **2024**, 146 (7), 4444.
3. da Hora, G. C. e. al. *J. Phys Chem B.* **2024**, accepted.

Designing Antimicrobial Coatings for Infection-Resistant Implantable Medical Devices

Anand Wadurkar, Matthew Libera, Annelise E. Barron, Shikha Nangia

Persistent infections on implantable medical devices are a significant healthcare challenge, often requiring multiple corrective surgeries, diminishing the quality of life in patients, and, in some cases, leading to amputations. This investigation explores a new class of peptidomimetic, biocompatible, synthetic, cationic polymers called peptoids with antimicrobial properties. When complexed with oppositely charged microgels, these cationic peptoids can serve as infection-resistant coatings on implantable devices to prevent infections. Our proof-of-concept study employs a synergistic computational and experimental approach to evaluate four peptoids for their efficacy as antimicrobial agents on titanium surfaces. Small-angle X-ray scattering data revealed that peptoids based on the building blocks, length, and sequence can have variable self-association properties ranging from small clusters to supramolecular assemblies. However, only the peptoids with smaller clusters demonstrated antimicrobial activity. We performed all-atom molecular dynamics simulations to understand the correlation between peptoid structure, assembly, and antimicrobial activity. Our results show a strong correlation between the chemical nature of the peptoid building blocks and their hydrophobicity. The tunability of peptoids has led us to a new project to examine the peptoid design space for optimal antimicrobial properties systematically. This collaborative project highlights the significance of computational design and experimental corroboration for engineering peptoid-based antimicrobial biomaterials to combat device-related infections.

Parch: A Robust Tool Quantifying Protein Hydropathy

Xuyang Qin, Jingjing Ji, Shikha Nangia

The quantitative assessment of the hydropathy of amino acid residues based on the nanoscale topography of the protein has been a longstanding research topic. In our recent work, we reported the development of a Protocol for Assigning a Residue's Character on the Hydropathy (PARCH) method that assigns the hydropathy of a residue on a 0 to 10 scale, where lower values indicate hydrophobic character. The PARCH scale considers both nanoscale topography and chemical properties of residues, which overcomes the limitations of the previous methods. In the original PARCH approach, we employed the three-point TIP3p water model that represents oxygen and hydrogen atoms optimized for compatibility with CHARMM36m proteins. However, to determine the robustness of the PARCH method, we examined three additional water models—TIP4p, TIP4pew, and TIP5p. Our findings reveal that while absolute hydropathy values may vary with different water models, the relative hydropathy between the residues on proteins was successfully reproduced regardless of water models. Notably, charged residues exhibit greater sensitivity to water model variations than others, showing larger value variances between different water models. We demonstrate that PARCH is robust in capturing relative hydropathy, independent of water model choice. It also facilitates PARCH's application across diverse protein-water forcefield combinations and even molecular systems beyond proteins.

Leveraging Machine Learning to Predict the Atmospheric Lifetime and the Global Warming Potential (GWP) of SF_6 Replacement Gases

Haewon Kim, Guobin Zhao, Yongchul G. Chung

The global warming potential (GWP) is a relative measure of the capability of a molecule to trap the Earth's infrared radiation as heat. The measurement or prediction of the GWP of a molecule is based on two factors: the radiative efficiency and the atmospheric lifetime of a molecule. While the calculation of the radiative efficiency of a molecule using the computational chemistry approach, such as density functional theory (DFT), is well-established and robust, the development of a computational approach to estimate the atmospheric lifetime remains challenging and limited to date. In this contribution, we developed a machine-learning approach to estimate a molecule's atmospheric lifetime and GWP100 based on electronic and geometrical features. We benchmarked the state-of-the-art computational workflow with the developed ML model in estimating the atmospheric lifetime and the GWP100. The developed ML model outperforms the existing approach with the MAE values of 0.234 (ML-predicted atmospheric lifetime), and 0.249 (direct ML model for GWP100) compared with 0.535 (Atkinson's method) and 0.773 (Kazakov et al.) from previous works. The developed models were used to screen >7,000 molecules in PubChem and bigQM7 datasets in a search for SF_6 replacement gas for the electric industry and identified 84 potential candidates.

Computational Discovery of Plastic-Binding Peptides for Remediating Microplastic Pollution

Michael Bergman, Carol Hall

Plastics on the micrometer or nanometer scale, termed microplastics or nanoplastics (MNPs), are a harmful pollutant. Millions of tons of MNP pollution pervade the environment. Humans and many other organisms regularly consume MNPs unknowingly, which can have harmful effects on their health. Because the risks of MNP to the environment and human health worsen with increasing MNP concentration, we urgently need methods to remediate MNP pollution. Here we discover plastic-binding peptides (PBPs) that could detect and capture MNP pollution by combining biophysical modeling, optimization techniques, and machine learning. PBPs were designed for polyethylene, polystyrene, polypropylene, and PET using the **Pep**tide **B**inding **D**esign (PepBD) algorithm. PepBD samples amino acid sequences and adsorbed conformations on a plastic surface using Metropolis Monte Carlo. Peptides were scored according to the sum of their MMGBSA binding energy and peptide stability in the adsorbed conformation (i.e., the peptide internal energy). Molecular dynamics (MD) simulations of designed PBPs show that they consistently have stronger binding free energies than random amino acid sequences, while preliminary atomic force microscopy experiments found that the best PBP for polyethylene has greater affinity than random sequences. An additional round of PBPs were designed to 1) increase PBP solubility in water, improving detection and capture of MNP pollution in aqueous environments, and 2) increase PBP binding affinity to a specific plastic, enabling separation or characterization of the components of MNP waste. Water solubility was increased by adding the CamSol solubility term into the peptide scoring function. Binding specificity was improved by first training a long short-term memory (LSTM) network to predict the PepBD score of an amino acid sequence, and then searching for sequences with large score differences between plastics. Experimental testing of PBP designs is ongoing. We envision that the PBP designs will be of great help in addressing MNP pollution.

Vapor–Liquid Equilibrium and Thermodynamic Properties of Saturated Argon and Krypton from Monte Carlo Simulations Using *Ab Initio* potentials

Philipp Ströker, Karsten Meier

Vapor–liquid equilibria and thermodynamic properties of saturated argon and krypton were calculated by semi-classical Monte Carlo simulations with the $NpT+$ test particle method using *ab initio* potentials for the two-body and nonadditive three-body interactions. The $NpT+$ test particle method was extended to the calculation of second-order thermodynamic properties, such as the isochoric and isobaric heat capacities or the speed of sound, of the saturated liquid and vapor by using our recently developed approach for the systematic calculation of arbitrary thermodynamic properties in the isothermal–isobaric ensemble. Generally, the results for all simulated properties agree well with experimental data and the current reference equations of state for argon and krypton. In particular, the results for the vapor pressure and for the density and speed of sound of the saturated liquid and vapor agree with the most accurate experimental data for both noble gases

almost within the uncertainty of these data, a level of agreement unprecedented for many-particle simulations. This study demonstrates that the vapor–liquid equilibrium and thermodynamic properties at saturation of a pure fluid can be predicted by Monte Carlo simulations with high accuracy when the intermolecular interactions are described by state-of-the-art *ab initio* pair and nonadditive three-body potentials and quantum effects are accounted for.

Optimization of a Monte Carlo Code for the Simulation of Thermodynamic Properties of Fluids Using Pair and Nonadditive Three-Body *Ab Initio* Potentials

Maohong Sui, Philipp Marienhagen, Johann A. Duffek, Philipp Neumann, Karsten Meier

In recent works, we calculated thermodynamic properties and the vapor-liquid equilibrium of pure nonpolar fluids using highly accurate *ab initio* potentials for the intermolecular interactions by Monte Carlo simulations. We showed that both thermodynamic properties and the vapor-liquid equilibrium can be predicted with uncertainties comparable to those of the most accurate experimental techniques if nonadditive three-body interactions and nuclear quantum effects are accounted for. Such simulations, especially, the evaluation of the nonadditive three-body potentials require large computational resources that scale with N^3, where N is the number of particles. To reduce the time-to-solution, we optimized our code, which is written in Fortran and parallelized with MPI, comprehensively.

Our simulations are performed in the NpT ensemble, in which the Markov chain consists of random particle displacements and volume changes because no derivatives of the potential with respect to particle distance are required in the calculation of thermodynamic properties. In the first step, a scalar optimization was carried out. Trigonometric function evaluations in the potential calculation were eliminated by using trigonometric identities and vector algebra. Moreover, the distance calculation for pair interactions was integrated into the distance calculation for three-body interactions. In the second step, we vectorized the distance and potential calculations using the AVX-512 instruction set and improved the load balancing in the potential calculation. For this purpose, the distance and potential calculations were separated into two loops. The outer loop of the three nested loops in the distance calculation for three-body interactions was distributed among the MPI ranks. First, the particle distances of all triplets are calculated and stored. In a subsequent single loop, the particle triplets inside the cutoff spheres are filtered by the cutoff criterion, and the distances of all filtered triplets are stored. In a further single loop, the potential is calculated for all filtered triplets. With these measures, the inner of the three nested loops and the single loops fully vectorize. A statistical analysis of the distribution of the filtered triplets among the MPI threads revealed large differences in the numbers of filtered triplets on each MPI thread, which resulted in considerable load imbalances among the threads in the potential calculations. These imbalances are much more severe for three-body interactions than for pair interactions. To reduce the time-to-solution further and to improve the scaling of the code, the load imbalances are reduced by transferring a part of the triplets from the threads with high load to those with low load by a point-to-point communication scheme before the potential calculation. With this strategy, we achieve an overall reduction of the time-to-solution by about a factor of eight to ten. The optimization was carried out using argon as an example, but our strategy can also be applied to Monte Carlo simulations of systems of molecules and molecular dynamics simulations.

Simulating Hydrodynamic Interactions in Soft Materials with Multiparticle Collision Dynamics

Jinny Cha, Chase Petix, Wilfred Kwabena Darko, Jeremy Palmer, Michael Howard

Understanding the behavior of soft materials, such as nanoparticles or polymers, suspended in a solvent, such as water, is critical for addressing numerous challenges, including improving the efficiency of wastewater treatment technologies, processing advanced materials for energy applications, and effectively delivering drugs to the body. Solvent-mediated hydrodynamic interactions typically play an important role in suspensions, but the solvent is difficult to model at process-relevant scales because of its significantly smaller size than the suspended materials. Multiparticle collision dynamics (MPCD) is a mesoscopic, particle-based solvent model that can flexibly address this challenge. Here, we present a new user interface to a massively parallel implementation of MPCD for both CPUs and GPUs that is publicly available in HOOMD-blue. We use our new software to critically assess how the orientation of a solid boundary relative to the MPCD collision-cell grid impacts accuracy of the method for a rotated parallel-plate geometry. The understanding we gain from this study will help us implement new algorithms for modeling complex solid boundaries such as those found in microfluidic devices and porous media using MPCD.

Electrically Active Induced Energy Levels and Metastability of B and N Vacancy-Complexes in 4H–SiC for Application in Quantum Technology

Emmanuel Igumbor, Edwin Mapasha

Electrically active induced energy levels in semiconductor devices could be beneficial to the discovery of an enhanced p or n-type semiconductor. Nitrogen (N) implanted into 4H–SiC is a high energy process that produced high defect concentrations which could be removed during dopant activation annealing. On the other hand, boron (B) substituted for silicon in SiC causes a reduction in the number of defects. This scenario leads to a decrease in the dielectric properties and induced deep donor and shallow acceptor levels. Complexes formed by the N, such as the nitrogen-vacancy centre, have been reported to play a significant role in the application of quantum bits. In this paper, results of charge states thermodynamic transition level of the N and B vacancy-complexes in 4H–SiC are presented. We explore complexes where substitutional N_{Si}/N_C or B_{Si}/B_C sits near a Si (V_{Si}) or C (V_C) vacancy to form vacancy-complexes (N_CV_C, N_CV_{Si}, $N_{Si}V_C$, $N_{Si}V_{Si}$, B_CV_C, B_CV_{Si}, $B_{Si}V_C$ and $B_{Si}V_{Si}$). The energies of formation of the N related vacancy-complexes showed the N_CV_C to be energetically stable close to the valence band maximum in its double positive charge state. The N_CV_{Si} is more energetically stable in the double negative charge state close to the conduction band minimum. The $N_{Si}V_C$ on the other hand, induced double donor level and the N_CV_{Si} induced a double acceptor level. For B related complexes, the B_CV_C and $B_{Si}V_C$ were energetically stable in their single positive charge state close to the valence band maximum. As the Fermi energy is varied across the band gap, the neutral and single negative charge states of the $B_{Si}V_C$ become more stable at different energy levels. B and N related complexes exhibited charge state controlled metastability behaviour. The results provided will provide a theoretical insight on defects-controlled phenomena for application in quantum technologies.

Efficient Path Integral Simulations Using Harmonic Oscillator Based Coordinates: Applications to Quantum Oscillators and Crystals

Sabry Moustafa, Andrew Schultz

Nuclear quantum effects play an important role in describing the behavior of lightweight molecules, especially at low temperatures. Path integral molecular dynamics (PIMD) and Monte Carlo (PIMC) simulations provide powerful tools to describe these effects. However, these simulations are usually computationally expensive due to the many replicas required for convergence. Nevertheless, the analytically known harmonic character of crystalline systems provides an opportunity to accelerate calculations. In this study, we employ the harmonic oscillator (HO) model to develop a novel framework aimed at enhancing sampling efficiency and improving the precision in the energy estimations. These achievements are grounded on introducing normal mode and staging coordinates which decouple the entire PI action of the HO model, rather than merely the ring-polymer kinetic part as normally done. We applied this approach to harmonic and anharmonic oscillators, and Lennard Jones crystal. In all systems, the HO-based coordinates results in high precision (and more move acceptance) and allowed for large timestep when applied to PIMC and PIMD, respectively, compared to standard ring-polymer based sampling methods. For the second task, we reformulate the conventional PI energy and heat capacity estimators using the recently developed harmonically mapped averaging (HMA) technique using the new coordinates. The results show a remarkable reduction in the uncertainty (hence shorter CPU time) when compared to the well-known centroid virial estimator. Although both normal mode and staging coordinates provide similar efficiency in terms of both sampling and estimator precision, an appealing feature of the latter is its linear scaling with the number of beads, rather than quadratic as in the former case. The synergistic impact of enhanced sampling and precise estimators obtained by this framework opens the door for efficient estimation of nuclear quantum effects for more complicated systems.

Theoretical Investigation of the Electronic. Energetic and Mechanical Properties of Polyvinyl Alcohol/Cellulose Composite Hydrogel Electrolyte

Nkechi Offia-Kalu, Abdulhakeem Bello, Simeon Nwanonenyi

The energy density of electrolytic hydrogel is influenced by many factors, such as water content, temperature, additives, and others, which impact its electrical, mechanical, and cohesive properties. This study examined the correlation between these properties using Density Functional Theory and molecular dynamics modeling techniques to analyze the quantum chemical descriptor of the core polymer. The binding energies of the hydrogel polymer composite increased as the quantity of water molecules in the composite increased to 60%, after which a decrease in the binding energy was observed. While, increasing the temperature led to a decrease in the binding energy of the composite. The highest cohesive energy density was observed with a water content of 40% and decreased as the temperature increased. The elastic characteristics were also influenced by an increased water content for up to 256 molecules. This result underscores the significance of quantum material science in the field of energy storage

Role of Hydropathy in Packing DNA Around Histone Proteins

Ratnakshi Mandal, Jingjing Ji, Shikha Nangia

Two meters of deoxyribonucleic acid (DNA) pack into a micron-sized nucleus to form the chromosomes. The DNA molecule, characterized by high negative charges and repulsive forces, cannot pack by itself, and must wrap around histone proteins, creating a complex protein-DNA core-shell structure. This DNA-protein complex is called a nucleosome, and this folds and assembles to form chromatin, which ultimately leads to the formation of a chromosome. Myriads of non-bonded interactions—electrostatic, van der Waal's, and hydrogen bonds—govern this assembly process. This study focuses on the hydropathic interactions associated with nucleosome formation using the recently developed PARCH method. PARCH is a unique hydropathy scale based on the chemical nature and the three-dimensional geometry of the molecules. PARCH method gives us *two* hydropathy values for every nucleotide, one for the backbone and one for the bases. For proteins, PARCH provides one hydropathy value for every amino acid. Using PARCH, we see that the backbone of the DNA is more hydrophilic than the nucleic bases, and the level of hydrophilicity of the backbone depends on the proximity of DNA to the core histones. The DNA regions proximal to histones, within 3Å cutoff, are hydrophobic, while the distal DNA regions are more hydrophilic. The histones also show varying levels of hydropathy, with the tails being more dynamic and hydrophilic while the core is stable and hydrophobic. The histone residues in contact with the DNA residues are hydrophobic, while they are more hydrophilic without the DNA. Our work shows that the assembly of DNA into chromosomes relies on hydropathy, which can be quantified using the PARCH method. This work is foundational to our ongoing study of the role of epigenetics on cancer.

Accelerating Scientific Discovery with HPC and AI

Nathan Baker

A new generation of artificial intelligence (AI) tools and cloud high-performance computing (HPC) is scaling and accelerating research and discovery in computational chemistry and materials science, thus unlocking opportunities for scientists to solve some of our world's toughest challenges. Tangible examples of experimentally validated computational discoveries are rare, but the number of such instances is beginning to increase as scientists employ these powerful new tools. In this talk, methods that accelerate discoveries in computational chemistry and materials science will be presented. Additionally, a recent case study involving the rapid discovery of a sustainable energy-storage solution will be discussed, in which new HPC and AI tools allowed scientists to discover, synthesize, and experimentally validate a novel material that functions as a low-lithium, solid-state battery electrolyte. This entire process, from discovery to experimental demonstration, spanned less than nine months. These recent advancements in HPC and AI are already accelerating scientific breakthroughs, with the goal of compressing the next 250 years of chemistry into the next 25. To add to the tools already available in Azure Quantum Elements, exciting new capabilities are coming soon, and will also be shared in this talk.

A Chemistry-Informed Machine Learning Framework for Closed-Loop Material Discovery

Paulette Clancy, Maitreyee Sharma Priyadarshini, Oluwaseun Romiluyi, Yirun Wang, Kumar Miskin

A major factor contributing to the inefficiency of material discovery tools is the large combinatorial space of materials which are often sparsely observed for a given application. Searches of this space are often biased by expert knowledge and clustered close to known material configurations. Moreover, exhaustive experimental characterization or first principles calculations are extremely expensive, leading to small available data sets. As a result, there is a need to develop algorithms that can efficiently search this large space. In this talk, we will introduce PAL 2.0, a method that combines a chemistry-informed belief model with Bayesian optimization. A key characteristic of PAL 2.0 is the creation of a physics-based hypothesis using XGBoost and Neural Networks which provide a physics-based prior to the Gaussian process model used in Bayesian optimization. Our method is unique since it picks out the physical descriptors that are most representative of the material domain, making the search unbiased toward expert knowledge. We demonstrate PAL 2.0's performance on three computational semiconductor data sets. We will also present how PAL 2.0 is being used in a closed-loop setup with experimentalists to discover high-temperature shape memory alloys for space actuation applications.

Advanced Sampling of Resilin-Like Peptides on 2D-Material Surfaces to Control Elastomer Cross-Linking

Tiffany R. Walsh

Cross-linked elastomers made from resilin, a highly elastic protein found in fleas and other insects, possesses remarkable mechanical and energy-storage properties that thus far cannot be recovered in the lab. These materials are produced via interchain cross-linking of tyrosines in the reslin-like peptide (RLP) domains of the protein. The spatial arrangement of the resilin chains in the cross-linked material exerts a strong influence over the mechanical properties in the organism. However, to date lab-based resilin-based materials cannot recover the natural elastomer's performance, due to a lack of knowledge regarding this spatial organization. The molecular-scale details of the RLP organization are challenging to obtain via experiment alone; molecular simulations have to potential to provide ground-breaking insights but have been under-utilized in this area. This presents the opportunity to shift resilin into the realm of multifunctional high-performance materials, via the discovery of new fundamentals regarding their structure/property relationships. Here, molecular simulation efforts are closely integrated with experiment to reveal the fundamental knowledge required to design and create 2D-material-templated RLP-based elastomeric films and fibers with precisely controlled structures and mechanical properties. This is achieved via advanced sampling of RLP and RLP-based conjugates with materials-binding peptides, identifying how and where materials-binding peptides can be linked to the RLPs. This work forges fundamental pathways to enable elastomeric resilin-based materials, including nanocomposites, with unprecedented properties.

Computational Chemical Separations

Jianwen Jiang

The energy usage of chemical separations in industries may constitute as high as 10–15% of the world's energy consumption, in which about 80% is consumed by thermally-driven energy intensive distillation. It is thus highly desirable to develop advanced technologies such as adsorption and membrane-based chemical separations. Toward the development of high-performance adsorbents and membranes from new materials, it is indispensable to develop quantitative structure-performance relationships and design guidelines from bottom up. With ever increasing computational power and resource, computation has become an indispensable tool and is playing an increasingly important role in screening and design of new materials for chemical separations.

This presentation will briefly summarize the recent computational (including molecular simulation and machine learning) efforts on adsorption and membrane-based chemical separations using both crystalline and amorphous materials. For crystalline materials, the focus is on metal-organic frameworks (MOFs) in many potential applications such as CO_2 capture, hydrocarbon separation and water desalination. For amorphous materials, polymer membranes will be discussed in CO_2 capture, biofuel purification and solvent recovery.

From chemical identity to Boltzmann ensembles for proteins, RNA and crystals with generative AI and statistical mechanics

Pratyush Tiwary

Modern structure prediction tools using Artificial Intelligence (AI) have made great progress in predicting single dominant structures for generic proteins, RNA and crystals. However, chemistry is often more than a single structure, and involves carefully predicting an ensemble of structures with right fluctuations and thermodynamic weights. Here I will discuss methods from my group that, through combining generative AI with statistical mechanics and molecular simulations, allow predicting an ensemble of atomic resolution conformations together with correct Boltzmann ranking. The examples I show will likely include protein kinases, human RNA and crystal polymorph nucleation, where we will obtain thermodynamic and dynamic observables such as drug residence times, conformational populations, effect of mutations, and melting curves starting solely from chemical identity and associated force-fields. I will conclude with thoughts on how chemistry, specifically classical thermodynamics and statistical mechanics have much to help design the next generation of AI methods meant for chemistry applications with all its quirks and richness.

AIMNet2: A Neural Network Potential to Meet Your Neutral, Charged, Organic, and Elemental-Organic Needs

Olexandr Isayev

Machine learned interatomic potentials (MLIPs) are reshaping computational chemistry practices because of their ability to drastically exceed the accuracy-length/time scale tradeoff. Despite this attraction, the benefits of such efficiency are only impactful when an MLIP uniquely enables insight into a target system or is broadly transferable outside of the training dataset, where models achieving the latter are seldom reported. In this work, we present the 2nd generation of our atoms-in-molecules neural network potential (AIMNet2), which is applicable to species composed of up to 14 chemical elements in both neutral and charged states, making it a valuable model for modeling the majority of non-metallic compounds. Using an exhaustive dataset of 20 million hybrid quantum chemical calculations, AIMNet2 combines ML-parameterized short-range and physics-based long-range terms to attain generalizability that reaches from simple organics to diverse molecules with "exotic" element-organic bonding. We show that AIMNet2 outperforms semi-empirical GFN-xTB and is on par with reference density functional theory for interaction energy contributions, conformer search tasks, torsion rotation profiles, and molecular-to-macromolecular geometry optimization. Overall, the demonstrated chemical coverage and computational efficiency of AIMNet2 is a significant step toward providing access to MLIPs that avoid the crucial limitation of curating additional quantum chemical data and retraining with each new application.

Accurate Free Energies of Aqueous Electrolyte Solutions from Molecular Simulations with Non-Polarizable Force Fields

Thijs J. H. Vlugt, Parsa Habibi, Othonas A. Moultos, Poulumi Dey

Efficient and accurate modeling of aqueous electrolyte solutions is a significant challenge. Non-polarizable force fields of aqueous electrolyte solutions are very popular for large scale molecular simulations due to their simplicity, accuracy, and transferability. The rigid non-polarizable water force fields of TIP4P/2005 (*Journal of Chemical Physics*, *123*.23 (2005), cited 3668 times) and SPC/E (*Journal of Physical Chemistry*, *91*.24 (1987), cited 14036 times) can accurately predict the structural properties, transport properties, densities, and interfacial tensions of liquid water. Many non-polarizable force fields of salts have been developed based on these water force fields. The charges of salt ions are commonly scaled down to account for the dielectric screening that occurs in the aqueous media. Using these scaled charge force fields the densities, dynamic viscosities, and self-diffusivities of aqueous electrolytes have been accurately predicted with respect to experiments. Despite this success, non-polarizable force fields fail to accurately capture free energies of the aqueous electrolyte solutions leading to highly inaccurate Vapor-Liquid Equilibria (VLE) predictions (e.g., the saturated vapor pressures of TIP4P/2005 at 300 K deviate by ca. a factor 4 compared to experiments). Non-polarizable water and ion force fields that accurately capture the free energies of the aqueous solution (e.g. TIP4P) struggle to predict the transport properties of the liquid phase (e.g. the viscosity of TIP4P at 298 K differs by ca. a factor 2 compared to experiments). Clearly, the transport properties and free energies of aqueous solutions cannot be modeled using a single available non-polarizable force field. We implemented a new approach in which (1) TIP4P/2005 water and scaled charge ion force fields are used to model the effective interactions in the aqueous phase, and (2) an additional Effective Charge Surface (ECS) is used to compute the free energies of water and ions with zero additional computational expense. A single temperature-independent charge scaling parameter, trained based on the excess chemical potential of water (with respect to the ideal gas reference state) at 350 K, is used to compute the ECS of water. This enables the accurate prediction of excess chemical potentials of liquid water at 300-500 K with an agreement of 1% compared to experiments. Based on a single ECS trained on the free energy of hydration of the scaled charge Madrid-2019 NaCl force field at infinite dilution, the free energy of hydration of different Madrid-2019 salt force fields (i.e., KCl, LiCl, $MgCl_2$, and $CaCl_2$) are accurately predicted with an agreement of 5% compared to experiments, compared to 20-30% deviations without the ECS. We devise a temperature-dependent free energy correction with which the VLE of water/salt mixtures at finite salt concentrations can be accurately predicted. Using this, the saturated vapor pressures of water/NaCl mixtures (until the solubility limit) at 350 K have been accurately described (within the error bars) compared to the experiments. This approach enables accurate free energy and VLE predictions using simple non-polarizable force fields, without compromising the thermophysical and transport properties of the aqueous phase.

Biomolecular Modeling and Simulation Today: Examples From Deciphering Mechanisms of Gene and Chromosomal Folding to Viral Frameshifting

Tamar Schlick

Molecular modeling and simulation is now recognized as a field on its own right capable of making key mechanistic insights into biomolecular processes. Notable successes include prediction of conformational transitions and mechanisms; generation of new insights into biomolecular activity; and thriving collaborations between modeling and experimentation, including experiments driven by modeling.

As examples from my lab, I I will describe mesoscale and multiscale modeling applications of genome folding, showing how we use tailored algorithms like DiSCO (Discrete Surface Charge Optimization) and Brownian dynamics to fold genes, recover chromosome conformations from Hi-C maps, and solve frontier problems in the epigenetic regulation of the genome and its relation to human disease. An entirely new set of methods, including graph theory and topology, enhanced sampling, statistical physics of conformational landscapes, and RNA coarse-grained folding and mutational predictions are involved in our pioneering work of developing RNA graphs to study viral frameshifting mechanisms. Such applications to the Covid-19 virus, SARS-Cov2, define new therapeutical avenues for viral suppression.

Structural Odyssey of Amyloid-β: Navigating the Conformational Landscape in Alzheimer's Pathogenesis

Birgit Strodel

Amyloid-β peptide (Aβ) plays a critical role in Alzheimer's disease, particularly in its neurotoxic, smaller oligomeric forms. Experimental challenges complicate the study of these transient structures, which has led us to utilize advanced molecular dynamics (MD) simulations and analysis techniques that include methods to calculate free energies and mean first passage time distributions to identify metastable states and frustrations in the energy landscape. This combination of methods applied to Aβ reveals the structural intricacies of the peptide and reveals its transformative journey from disorder to order through interactions with other biomolecules, including other Aβ peptides. In particular, a stable β-hairpin motif emerges during self-assembly that favors ordered oligomers with a higher propensity to aggregate. Exploring the conformational landscape of Aβ improves our understanding of its dynamics and provides insights that are critical for therapeutic interventions in Alzheimer's disease.

Multiscale Computational Models for Uncovering the Molecular Language of Biomolecular Condensates

Jeetain Mittal

Biomolecular condensates formed through phase separation offer intriguing insights into fundamental questions surrounding cellular organization. This phenomenon is suggested to play a role in various biological processes, including chromatin organization, DNA repair, transcription, and gene regulation. The aberrant phase separation of functional proteins has been linked to cellular stress, aging, and diseases such as cancer and neurodegenerative disorders, including Alzheimer's disease, Parkinson's disease, amyotrophic lateral sclerosis (ALS), and frontotemporal dementia (FTD). Conversely, the harnessing of underlying principles in macromolecular phase separation presents an opportunity to engineer intracellular compartments with novel functionalities. The long-term focus of my group is to decipher the relationships between amino acid sequences, phase behavior, condensate dynamics, and biological function. In this talk, I will describe our recent work, combining physics-based simulations and experimental techniques, to uncover the physicochemical principles of biomolecular phase separation, offering both general understanding and system-specific insights.

Molecular Modeling of Nucleation of Molecular Crystals

Erik Santiso

Crystallization is a process that is common in chemistry and chemical engineering as a means to purify compounds, or to prepare products in solid form. It has applications in the manufacturing of some food products, specialty chemicals, energetic materials, and pharmaceuticals, among others. Yet a detailed, molecular-level understanding of the crystallization of molecular crystals, and particularly the initial stages of crystal nucleation, is difficult to obtain. Nucleation is a stochastic process that often involves a small number of molecules, and it is affected dramatically by minute molecular-level effects, including the presence of small impurities or defects on surfaces, making its experimental study challenging. On the other hand, the time scales for nucleation can be of the order of minutes, hours, or more, which puts its study beyond the limit of traditional molecular dynamics simulations and requires the use of rare event techniques.

We have developed a general method to construct order parameters (OPs) sensitive to the structure of a given molecular crystal. By combining these OPs with the String Method in Collective Variables (L. Maragliano et al., *J. Chem. Phys.* **125**, 024106, 2006), we have obtained nucleation pathways for different organic molecules from the melt and from solution, including some active pharmaceutical compounds. More recently, we have extended the method to generate OPs that are sensitive to the local structure of liquid solvents, which can be used to investigate the effect of interactions between the solvent and the incipient crystal on the nucleation process. In this talk we will present an overview of our approach to different systems, including a discussion of solvent effects, and discuss current plans for extending our approach to study more complex systems.

Self-Assembly Models for Crystal Growth and Phase Transitions

Julia Dshemuchadse

How can we make new materials and better understand how their underlying structures form? The direct observation of crystal growth and transitions remains supremely challenging, but gaining insight into these fundamental processes is central to our quest of creating materials in a rational and targeted way, connecting structure to functionality. We build self-assembly models, study how they react to perturbations on the particle and system levels, and investigate their impact on crystal growth and transformation pathways. We use simple coarse-grained models to gain systematic insights into the phenomena that lead to the crystallization of complex crystal structures, partial disorder, or magic-size assemblies, allowing us to derive the essential principles that govern the formation of materials' structures. Our goal is to use these insights to find ways to tailor crystallization pathways and to create new functional materials. Our work promises to establish new pathways to materials design through simulations, which explicitly incorporate and explore phase transformation kinetics.

Beyond Conventional Transition State Theory in Catalysis: Applications of Matrix Completion

Shaama Mallikarjun Sharada

Chemical reactions lie at the heart of processes designed to meet our growing energy and material needs. The first step towards designing and optimizing chemical reactions involves identification of underlying mechanisms and quantification of rates. Quantum chemistry methods along with theories such as transition state theory (TST) are indispensable for this purpose and have played a pivotal role in elucidating mechanisms in recent decades. While widely successful, conventional TST is relatively simplistic and can lead to inaccurate rates for many classes of reactions. Alternative, more accurate rate theories such as variational transition state theory with multidimensional tunneling (VTST-MT) are well-established but incur exceptionally high computational costs which limits their widespread use. We aim to lower these costs to enhance reliability of rate predictions by adapting algorithms typically used in signal processing. I will present our algorithm – polynomial variety-based matrix completion (PVMC) - that leverages matrix completion methods, widely used to recover signals from noisy, incomplete data, to recover otherwise expensive second derivatives of energy for points on the minimum energy path of a reaction. The algorithm is up to an order of magnitude cheaper than traditional VTST approaches, and we demonstrate its utility in practical problems in catalysis with enzyme-inspired CH activation chemistry.

Influence of Liquid Water on the Thermodynamics, Kinetics, Mechanism, and Rate of Hydrogen Production from Methanol over Pt/Al2O3 Catalysts

Ricardo García Cárcamo, Tianjun Xie, Bryan Hare, Carsten Sievers, Rachel Getman

One of our most important scientific challenges is to design catalysts that produce H_2 from "minimum CO_2" sources. One strategy is by aqueous phase reforming (APR) of sugar alcohol molecules derived from biomass. However, H_2 yields have been disappointing, indicating that catalysts and reaction conditions require optimization. This requires a detailed understanding of the mechanism. There are three primary steps: dehydrogenation, decarbonylation, and water gas shift. However, details of the dehydrogenation step remains unknown due to the large and complex structures of the reactant molecules, the aqueous reaction conditions, and the participation of multiple active sites in the mechanism. To address these things, we study the effect of liquid H_2O solvent and multiple active sites on the mechanism of the APR of CH_3OH. Specifically, we compare microkinetic modeling results with observations from ATR-IR to provide insight into the dehydrogenation of CH_3OH on Pt/Al_2O_3. Microkinetic modeling utilizes interfacial free energies calculated with our method of multiscale sampling, which combines DFT with classical MD to address explicit influences of solvent molecules on interfacial species enthalpies and entropies. Further, we investigate sites on the terraces of large Pt particles as well as at the Pt/Al_2O_3 perimeter. We show that while Pt terraces follow an "alcohol route", the Pt/Al_2O_3 perimeter follows an "aldehyde route". Despite a small promotional effect in aiding to break the CO-H bond, liquid H_2O ultimately inhibits the rate in both sites. On terrace sites, this is due to stabilization of COH, which impedes dehydrogenation. On perimeter sites, H_2O binds strongly to Al_2O_3, inhibiting CH_3OH diffusion to the Pt/Al_2O_3 perimeter. Further, it donates protons to CH_xO species, effectively pushing the reaction in reverse. Understanding of these details enables envisioning ways to tune catalyst composition and solvent effects to better promote the rate of sugar alcohol dehydrogenation on supported metal catalysts.

AUTHOR INDEX

AIChE
120 Wall Street, FL 23
New York, NY 10005-4020

ISBN 979-8-3313-1718-8
9 798331 317188

CURRAN ASSOCIATES INC.
proceedings.com